PLUTONIUM

A HISTORY OF THE WORLD'S MOST DANGEROUS ELEMENT

PLUTONIUM

A HISTORY OF THE WORLD'S MOST DANGEROUS ELEMENT

Jeremy Bernstein

Cornell University Press

ITHACA AND LONDON

First printing, Cornell Paperbacks, 2009

Printed in the United States of America

Library of Congress Cataloging-in-Publication Data
Bernstein, Jeremy, 1929–
   Plutonium: a history of the world's most dangerous element / by Jeremy
   Bernstein.
   p. cm.
   Includes index.
   ISBN 978-0-8014-7517-7
   1. Plutonium—History. I. Title.
   QD181.P9B47 2007
   546'.434—dc22

Cover design by Michele de la Menardiere

Cornell University Press strives to use environmentally responsible suppliers and materials to the fullest extent possible in the publishing of its books. Such materials include vegetable-based, low-VOC inks and acid-free papers that are recycled, totally chlorine-free, or partly composed of nonwood fibers. For further information, visit our website at www.cornellpress.cornell.edu.

Paperback printing   10 9 8 7 6 5 4

Contents

Acknowledgments

When I began writing books, now some decades ago, they were made up of things that had first appeared in the *New Yorker*. My acknowledgments were to *New Yorker* editors and above all to William Shawn, who had first encouraged me to write about science for the general public. I also acknowledged the *New Yorker* checkers who kept me from making all sorts of factual mistakes. And I thanked my first great teacher in physics—Philipp Frank—who inspired me to learn the subject. When I started writing books that did not appear first in the *New Yorker,* they were about subjects about which I had been thinking for many years, the most recent example being a profile of Robert Oppenheimer. Here I acknowledged mainly people who had been encouraging, or sometimes critical, but did not really check the factual material, which I felt that I had pretty well in hand.

This book has been quite different. When I started it I knew rather little about plutonium. It is true that I had taught courses in nuclear physics, but in those courses plutonium enters as part of a lecture about fission. I never had occasion, for example, to learn about plutonium's totally bizarre chemical and physical properties to say nothing of its history. All of this I have learned in the course of writing this book. But I have had wonderful teachers who spent a great deal of time patiently explaining things to me and correcting mistakes. It is to them that I dedicate the book and make the following acknowledgments. I hope that I have not left anyone out. So I thank Elihu Abrahams, Steve Adler, Lorna Arnold, Andrew Brown, Norman Dombey, Freeman Dyson, Sheldon Glashow, Sig Hecker, Roald Hoffman, Roman Jackiw, Rainer Karlsch, Peter Kaus, Harry Lustig, Helmut Rechenberg, Oliver Sacks, Carey Sublette, Erick Weinberg, and Peter Zimmerman. I also thank Jeffrey Robbins, Lara Andersen, Christine Hauser, and Sally Stanfield for their work in bringing this book to its final state. Many thanks. As for all our work, as the Italians say, *Se sono rose, fioriranno.*

Jeremy Bernstein
Aspen, Colorado
August 10, 2006

Prologue

The origins of this book are somewhat idiosyncratic. In the spring of 2005, the German historian Rainer Karlsch published a book entitled *Hitler's Bombe*[1] that created something of a sensation. Karlsch found evidence that, all during the war, the Germans had a nuclear weapons program that they managed to conceal not only from the Allies but even from the people such as Werner Heisenberg who were working on what was acknowledged as the German nuclear energy program. In the spring of 1945, many of these German scientists, including Heisenberg, were captured, and 10 of them were interned by the British in a country estate, Farm Hall, near Cambridge. British Intelligence secretly recorded their conversations during the six months of their confinement.[2] Among the detainees were two physicists, Kurt Diebner, who had been a member of the Nazi party, and Walther Gerlach, both of whom apparently were deeply involved in the weapons program. They hid this fact not only from their interrogators but also from their fellow inmates. Some of the things that they were recorded as saying, especially their denial of the existence of such a program, were outright falsehoods.

That there was such a program does seem to be supported by the evidence. But what made Karlsch's book a sensation was his claim that these people, in the spring of 1945, in the woods in the

German province of Thüringen, set off one, or possibly two, nuclear explosions. While these explosions are not claimed to have been from "bombs," any report of a test by the Germans involving explosions and nuclear reactions was profoundly shocking. My own view, based on reading Karlsch and discussing the matter with a number of expert colleagues, is that although Karlsch has provided interesting new evidence for the existence of such a program, the notion that these explosions, if they took place at all, were nuclear is totally unsubstantiated. In fact, recent evidence shows that they could not have been nuclear. Indeed, what interested me about Karlsch's book was not this, but rather the various bits of documentary evidence he has discovered that stands on its own. Among the items is a 1941 patent application by Carl Friedrich von Weizsäcker, also a Farm Hall detainee, for the use of plutonium to make nuclear explosives by producing it in a reactor.[3] For reasons that are made clear later, Weizsäcker does not call the element in question plutonium. He refers to it as element "94," which is its place in the periodic table of elements. In other German wartime publications it is called "ekaosmium." Later, I explain where this odd name comes from. In any event, as I was reading Weizsäcker's patent application, the thought kept occurring to me that in light of the work that was going on in the United States and elsewhere, it was absurd. Later I provide much more detail, but let me give the outline here.

Karlsch does not give the precise date of Weizsäcker's filing, but it was in the spring or summer of 1941. It must be noted that at that time, and indeed at no time during the war, did the Germans produce any plutonium at all. (I am ignoring some claims of Karlsch, which seem to me to have no plausibility, that involve the use of a small cyclotron in Paris. No evidence has been produced that this cyclotron, which was in the hands of anti-Nazis, produced plutonium.) In contrast, in February of 1941, the nuclear chemist Glenn Seaborg and some colleagues, using a cyclotron at Berkeley, had succeeded in producing and identifying some micrograms of plutonium. Indeed, they had submitted a brief paper with their findings to the *Physical Review*, having named the element "plutonium." The paper carried the instructions, that because of its implications that plutonium was a possible nuclear explosive, it was not to be published until after the war. Indeed, the name was so secret that, all during the war, Manhattan Project scientists referred to it as "49." Later, I also explain how this curious name was arrived at.

By the end of March 1941, the experimenters in Berkeley, again using micrograms, had confirmed that plutonium was fissionable and therefore a potential ingredient for a nuclear weapon. The Germans, having no plutonium, simply assumed that it was fissionable on theoretical grounds. By later that spring, Seaborg and his colleagues had left Berkeley for Chicago, where they became part of the so-called Metallurgical Laboratory. Their first job was to learn how to separate plutonium from the uranium matrix in which it was created. This task turned out to be very difficult, even on a laboratory scale, let alone in the industrial context needed to produce kilograms of the stuff. In various publications during the war the Germans discussed such a separation in the abstract but never provided any specifics. In fact, what Weizsäcker writes about this in his patent application is entirely incorrect. He notes, ". . . the product 94 is easily separated from uranium (using the well known rules for ekarhenium or, respectively ekaosmium, or similar rules) and can be purified."[4] As discussed later, "ekarhenium" and "ekaosmium" were supposed to have the same chemical properties as rhenium and osmium, respectively. But both elements 93 and 94 turned out to have entirely different chemical properties, which is one of the reasons element 94 was not "easily separated from uranium." Since the Germans never produced any, they had no idea of the chemical complexity of plutonium. It is in many ways the most complex element there is. For example, its crystal structure changes five times as it is heated to its melting point. Sorting out all of this was a horrendously complicated task.

In view of all of this, Weizsäcker's patent application appears to me to be totally naïve, but it got me started thinking about the subject of plutonium, about which I realized I knew rather little.

I was hoping to find a book that would teach me. However, after searching the literature, I came to the conclusion that there isn't one. There are specialized monographs usually written for professionals, but no one has written about the history and science of plutonium, and its role in nuclear weapons, in an accessible form. This is what I have set out to do.

PLUTONIUM

A HISTORY OF THE WORLD'S MOST DANGEROUS ELEMENT

I
Preamble

spent the summer of 1957 as an intern at Los Alamos. This was the height of the Cold War and the laboratory was trying to recruit young scientists. I had no desire to make a career working at a weapons laboratory, but my curiosity got the better of me so I signed up for the summer. I think the place still resembled what it had looked like during the war. It was a closed city surrounded by guarded fences and I lived in one of the barracks that had probably housed a future Nobel Prize winner in wartime days. Although I had the required Q clearance, I was never privy to the work that must have been going on all around me on the design of nuclear weapons. I had made friends with a more senior colleague who was also there for the summer. We played tennis on a regular basis. One day he told me that he would not be able to have our usual game because he was going to Mercury, Nevada, to watch some aboveground atomic bomb tests. He said he was going because he was curious. I asked how one could arrange this, and he told me that I would have to speak to the head of the Theoretical Division, a Canadian named Carson Mark. I then asked Mark, and he said it would be fine with him providing I paid my own way, airplane fare and the like. And so it happened that the three of us went to see the tests.

I won't try to describe the impression that the actual explosion made on me except to say that it was a sight I have never gotten over. But let me describe what happened after we had seen the first test.

Mark took us on a tour of the site. We passed areas, where previous tests had taken place, that were still radioactive. We climbed a tower where the next weapon to be tested was being assembled. Then we went to a blockhouse that was well separated from the rest of the site. When I went inside, my heart stopped. On shelves were the "pits" of atomic bombs. These are the spheres at the center of the bombs that are made of plutonium shells covered with a nickel coating. Around these shells high explosive is glued; setting off the explosive simultaneously at several points on the sphere is what causes the plutonium to implode and starts the chain reaction. For whatever reason, maybe to see my reaction, Mark gave me a pit to hold. It was somewhat warm from the radioactive plutonium and about the size and weight of a bowling ball. My only thought was that I was holding the working interior of an atomic bomb and probably should not drop it. It did not occur to me at the time what a remarkable thing this was. I do not know the actual amount of plutonium I was holding. Several pounds worth. The plutonium was certainly in a thin shell. The kind of plutonium used in a nuclear weapon has a density twice that of wrought iron. A solid sphere of radius six inches would weigh about 500 pounds. Nonetheless, I was holding the several pounds of plutonium in one bomb. To see how remarkable this was, note that when what was then the entire world's supply of plutonium was weighed on September 10, 1942, in the Metallurgical Laboratory at the University of Chicago, it weighed 2.7 millionths of a gram!

II
The History of Uranium

The history of plutonium begins with the discovery of uranium.[1] In the sixteenth century, silver was found in a river in a mountainous region near Saxony in Germany. Because of the silver boom, a town was created for the miners that came to be called Sankt Joachimsthal and several silver mines were opened. While the silver boom ebbed and flowed, mining continued into the eighteenth century. Among other things the miners encountered was a shiny black mineral that they called pechblende—pitch mineral—pitchblende. It was first analyzed by a self-educated chemist named Martin Klaproth, and in 1789, he found in it what he called a "strange kind of half metal" that seemed to be a new element. Klaproth had no way of knowing that what he had discovered was the heaviest naturally occurring element. At first he was going to name it after himself, but on a tentative basis he decided to name it after the planet Uranus, which had been discovered by his countryman William Herschel in 1781. He admired Herschel and to honor him called the element uran, which later became "uranium."

In the years that followed, uranium was discovered in other locations, but nothing of scientific interest occurred until 1869, when the Russian chemist and general polymath Dimitri Mendeleev organized the 63 elements then known into what is now called the "periodic

3

table." To enable me to introduce a number of concepts that we need later, I am going to give the modern definition of "atomic weight," which was the organizing principle that Mendeleev used. Mendeleev did not know any of this. The atomic nucleus consists of electrically neutral and positively charged massive elementary particles called neutrons and protons, respectively. The neutron is slightly more massive than the proton. The total number of protons in a nucleus is called the "atomic number." In an electrically neutral atom, a corresponding number of negatively charged electrons circulate at a relatively great distance from the nucleus. These electrons determine the chemical behavior of the atom. Most elements are found in forms that have different numbers of neutrons but the same number of protons. These different forms are called "isotopes." To take one example, ordinary hydrogen has a nucleus with one proton. The nucleus of heavy hydrogen—deuterium—consists of one proton and one neutron. The usual notation for these two isotopes is ^{1}H and ^{2}H.

Very roughly, the mass of an atom is the sum of the masses of the neutrons and protons in its nucleus. This is an approximation because it ignores the masses of the light electrons and it ignores a result of Einstein's formula $E = mc^2$. When neutrons and protons combine to form a nucleus there is a mass loss. The whole is less massive than the sum of its parts. This loss is referred to as the "mass defect." To separate the nucleus into its constituent neutrons and protons, we must supply an amount of energy related to this mass loss by Einstein's formula. This mass loss is much less than the masses of the constituent neutrons and protons. As I have mentioned, elements are usually found with different isotopes that have different masses. These isotopes occur naturally in varying relative amounts. For example, most naturally occurring uranium occurs in the isotope uranium-238 whose nucleus consists of 92 protons and 146 neutrons. But there is an isotope uranium-235, which occurs in less than 1 percent of natural uranium, that has the same number of protons but three fewer neutrons. The atomic mass of an element is defined by taking the mass of each of its isotopes and averaging this with the frequency of its occurrence.[2] The atomic masses of elements are close to integers since they are nearly the sum of the neutron and proton masses of the most frequently occurring isotope. It was this atomic mass that, without an understanding of the underlying structure of the atom, Mendeleev used to organize his table.

To a modern reader, Mendeleev's first periodic table, which he published in 1869, looks rather odd.[3] In the first place, a modern periodic table is organized in terms of atomic numbers, not atomic weights. This is because we understand that the atomic number, which reflects the number of the electrons in the atom, is what determines its chemical characteristics. In the second place, Mendeleev put elements that have similar chemical properties in horizontal rows according to their increasing atomic weights. The vertical columns are the elements again arranged in terms of atomic weights. The periodicity comes in because the same chemical behavior repeats itself as the elements become heavier. Curiously, in this first periodic table Mendeleev does not list uranium. The heaviest element he has is lead, to which he assigns an atomic mass of 207, compared to the atomic mass of hydrogen, to which he assigns a value of 1. He did not know anything about the relatively rare isotopes of hydrogen that were only isolated in the twentieth century.

Mendeleev's second version of the periodic table, which he published in 1872, looks more familiar to a modern reader (see Figure 1). I have kept the original German used by Mendeleev. The elements with similar chemical properties are now arranged vertically so, for example, there is a vertical column that contains hydrogen, lithium, sodium, potassium, and so forth. He still uses atomic weights as his organizing principle, claiming, incorrectly, that these are what determine the chemistry. Uranium now appears in the table. With its atomic mass of 240, it is the heaviest element. In the same series he includes thorium with an atomic mass of 231. We now know that thorium is the lightest element in the so-called actinide series that, besides uranium, includes elements such as protactinium and

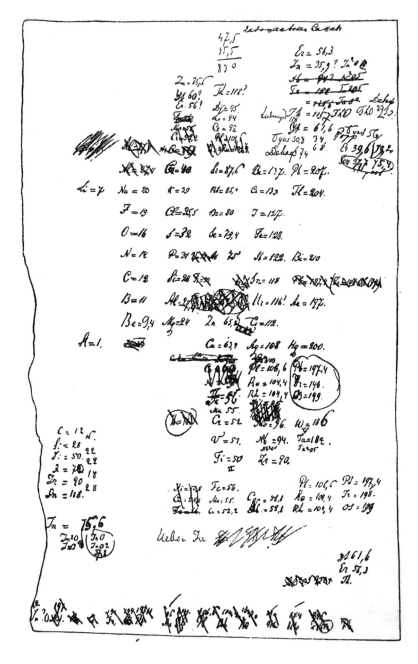

FIGURE 1 Mendeleev's 1872 periodic table, in his own hand, courtesy of Roald Hoffmann.

the transuranics such as plutonium, none of which were known to Mendeleev. Others had noticed these chemical repetitions, but what immortalized him was the use he made of them. He predicted, with great precision, "missing" elements.

If Mendeleev had organized his table in terms of atomic numbers, the location of the missing elements would have been rather transparent. There would not have been an element associated with one of the integers, the atomic numbers. Something would evidently have been missing. But having made the organization in terms of atomic masses, the problem was more subtle. He found jumps in the atomic masses that looked unnatural. Looking across the row, for example, which includes potassium (39) and calcium (40)—the atomic masses—he noticed that the next element in the table was titanium to which he assigned an atomic mass of 48. In a later paper he revised this number to 50. He was right the first time: The modern value is 47.9. In any event, Mendeleev predicted that there should be an element in between, to which he assigned an atomic mass of about 45. But this element, if it existed, fell into the third vertical column, so its chemistry could be predicted from the chemical behavior of the other elements in the column. It is amazing how accurately he predicted the detailed chemical properties of this element. Typically, he writes, "It should not decompose water at ordinary temperature but at somewhat raised temperatures . . . "[4] and so on. In 1878, the Swedish chemist Lars Frederic Nilson found a new element that he named scandium. It was later pointed out that this element, which has an atomic mass of 44.96, was just what Mendeleev had predicted. This confirmation of his prediction is what immortalized Mendeleev.

There is an odd twist to this history that has echoes in our plutonium story. Mendeleev became a professor of chemistry at the university in St. Petersburg, where his colleague Otto Böhtlingk taught Sanskrit. Mendeleev became interested in this ancient language, and although it is not clear how much he absorbed, he certainly learned some of its number system: for example, "*eka,*" "*dvi,*" "*tri*" for one, two, and three.[5] In naming the missing elements he employed these Sanskrit numbers. Of what was later named scandium he writes, "I have decided to give this element the preliminary name of *ekaboron*, deriving the name from this, that it follows boron as the first element of the even group, and the syllable *eka* comes from the Sanskrit word meaning 'one'."[6] What he has done is to go across the row in his table containing boron and to note that the missing element is one element over, hence *eka*. Likewise in his notation *eka-aluminum* was later called gallium. In the *dvi* category, *dvitellurium* later became polonium, and in the *tri* category, *trimanganese* became rhenium. This is somewhat puzzling because in both the modern and the Mendeleev periodic tables rhenium is two places over.[7] What is equally strange is, as we shall see, that this odd labeling in a somewhat different form persisted into the twentieth century and enters into our plutonium story. I wonder if any of the more recent scientists who used it knew its origin.

III
The Periodic Table

I want now to interrupt the historical narrative and describe briefly the modern view of the periodic table. This is a rich and wonderful subject whose surface we can only scratch. To explain the periodic table we must first have a correct model of the atom. The road to this model began in a series of experiments carried out at Manchester University by the New Zealand physicist Ernest Rutherford and his two junior colleagues Hans Geiger and Ernest Marsden. The object of their work, which began in 1909 and culminated in Rutherford's magisterial paper of 1911, was to scatter—that is, bounce—so-called alpha particles (which later turned out to be helium nuclei) from thin foils such as gold. The general expectation was that these particles, which were emitted from a radioactive source, would pass through the foils with little deflection. Nevertheless, on a hunch, Rutherford asked his young colleagues to look for scatterings at large angles. Much to his surprise, they found many more than Rutherford had expected. Rutherford explained these large-angle events with the notion that the alpha particles had hit something hard in the gold atoms. An image that is useful is to imagine firing a bullet into a bale of cotton in which a miniscule hard object has been hidden, in this case, the atomic nucleus within the atom. Rutherford used the image of firing a cannon ball into a tissue and having it bounce back. The question was, What was this nucleus composed of?

It was well known that the lightest nucleus was that of hydrogen, which has a positive charge that balances the negative charge of the electron circulating in some fashion around it. There were suggestions preceding Rutherford's discovery that the hydrogen nucleus might be a fundamental building block of matter. Rutherford agreed, and in a paper in 1920 he called it the "proton," but it was clear from the atomic masses that there must be something else in the nucleus. For example, the next heaviest element, helium, has a positive charge of two but an atomic mass of four. Whatever made up this mass had to be electrically neutral so as not to disturb the electrical neutrality of the atom. Rutherford made the natural guess that this mass was an object composed of a proton and an electron bound together. When, in 1932, the British physicist James Chadwick observed the effects of this neutral particle directly by bombarding beryllium with alpha particles, this is what he thought he had discovered. There were a few dissenters, notably the Austrian-born physicist Wolfgang Pauli, whose work we encounter shortly, who said that the neutral particle—the neutron—could not be such a combination but must be an elementary particle in its own right. This meant that there were no electrons in the nucleus, but what about those outside it?

The outside electrons posed two related problems. First there was the stability of the atom. The negatively charged electrons are attracted to the positively charged protons. Why don't they simply crash into the nucleus? The second problem had to do with atomic spectra. When excited, atoms produce beautiful patterns of radiation—some of which are visible. These patterns can be used to identify a particular kind of atom, and this is how many elements were discovered. But if the electrons that produce this radiation crash into the nucleus, why would this produce orderly patterns? An image that comes to mind is pushing a grand piano out a window and expecting it to emit Beethoven's "Moonlight Sonata" when it hits the ground. The beginnings of the solution to both of these problems were pioneered by Niels Bohr. Bohr had come from Denmark in 1911, after having spent an unsatisfactory year in Cambridge, to study with Rutherford in Manchester. He was 27 and extremely shy and reserved. Fortunately, despite this shyness, Rutherford soon recognized Bohr's extraordinary potential as a scientist. The two of them must have made quite a pair: Rutherford, who spoke in a booming voice, had no self-doubts at all when it came to physics. One night, one of his young associates awoke him at home with a telephone call about an experimental result that seemed strange. Rutherford produced, without much justification, an explanation. The young man, who did not quite understand, asked Rutherford his reasons. "Reasons! Reasons!" Rutherford bellowed on the phone. "I feel it in my water."[1] Bohr came to love Rutherford and kept up a relationship with him that lasted for the rest of Rutherford's life. On his return to Denmark in 1913, Bohr created his atom—and ours.

Upon returning to Denmark, Bohr learned of some theoretical work on the spectrum of hydrogen that had been done by a Swiss high school teacher—actually, he had a Ph.D. in mathematics—named Johann Jakob Balmer. There were at the time only four spectral lines that Balmer knew about. He realized that the frequencies—the colors—of these lines obeyed a simple mathematical law. They were given by a relationship that was proportional to the difference $1/n^2 - 1/m^2$, where n and m were integers. In particular, if Balmer took n to be 2 and let m be 3, 4, 5, 6, the pattern of the observed lines was fitted, and this pattern persisted when more lines were found. This was the clue Bohr needed. His idea was as follows.

The electron that circulated around the proton in the hydrogen nucleus could not have arbitrary energies; there were only certain allowed energies. The quantum of radiation out of which the observed spectrum was composed had a frequency that was proportional to the difference of a pair of these energies. The picture was that the electron jumped from an orbit with a higher energy to one with a lower energy and emitted a radiation quantum. Moreover, the state of lowest energy, usually called the "ground state," was stable against any further emissions. There was no place for an electron in it to go and lose more energy. The problem then was how to determine these allowed energies.

Bohr realized that the key to this problem's solution was the quantization of angular momentum. An object in a curving motion, such as the Moon going in a circle around the Earth, has a momentum that reflects this motion. For the Moon, to take an example, this momentum is proportional to the speed of the Moon in its orbit multiplied by its distance from the Earth. In classical physics there are no restrictions on the magnitude that this angular momentum can have. Likewise, a classical electron moving in a circular motion around a proton can have any angular momentum. Bohr made the radical assumption that, in fact, this angular momentum was proportional to some integer varying from zero to infinity. The zero angular momentum state was the ground state. He combined this assumption with the additional assumption that, apart from this condition, the motion of the electron was governed by the classical Newtonian laws of motion. Clearly this was an uneasy mixture of assumptions but, using them, Bohr readily showed that the electron could move only in restricted orbits characterized by radii that were proportional to the square of the same integer that quantized the angular momentum, the so-called Bohr orbits. Once he knew how these radii depended on this integer, it was easy to derive how the energy of these orbits depended on the integer. Indeed, it depended as $1/n^2$, just as Balmer's formula demanded. In fact, the proportionality factor that Bohr derived for the frequency of the lines was in excellent agreement with the same constant Balmer had gotten from experiment. Clearly, a corner of the veil had been lifted. For the next decade this mixture of classical and quantum assumptions was developed in various directions, some successful and many not. The whole enterprise fell under the rubric of the "old quantum theory." One of its activities was to try to understand the periodic table.

Figure 2 is a modern version of the periodic table. Clearly it is a very far cry from Mendeleev's. It is organized in terms of the

1 H																	2 He
3 Li	4 Be											5 B	6 C	7 N	8 O	9 F	10 Ne
11 Na	12 Mg											13 Al	14 Si	15 P	16 S	17 Cl	18 Ar
19 K	20 Ca	21 Sc	22 Ti	23 V	24 Cr	25 Mn	26 Fe	27 Co	28 Ni	29 Cu	30 Zn	31 Ga	32 Ge	33 As	34 Se	35 Br	36 Kr
37 Rb	38 Sr	39 Y	40 Zr	41 Nb	42 Mo	43 Tc	44 Ru	45 Rh	46 Pd	47 Ag	48 Cd	49 In	50 Sn	51 Sb	52 Te	53 I	54 Xe
55 Cs	56 Ba	57 La	72 Hf	73 Ta	74 W	75 Re	76 Os	77 Ir	78 Pt	79 Au	80 Hg	81 Tl	82 Pb	83 Bi	84 Po	85 At	86 Rn
87 Fr	88 Ra	89 Ac	104 Rf	105 Ha	106 Sg	107 Ns	108 Hs	109 Mt	110	111	112	(113)	(114)	(115)	(116)	(117)	(118)

LANTHANIDES

58 Ce	59 Pr	60 Nd	61 Pm	62 Sm	63 Eu	64 Gd	65 Tb	66 Dy	67 Ho	68 Er	69 Tm	70 Yb	71 Lu

ACTINIDES

90 Th	91 Pa	92 U	93 Np	94 Pu	95 Am	96 Cm	97 Bk	98 Cf	99 Es	100 Fm	101 Md	102 No	103 Lr

FIGURE 2 The modern periodic table.

atomic number—the number of electrons circulating around the nucleus—and the families are in vertical columns rather than rows. Mendeleev's missing elements have been filled in. But it is also a far cry from the periodic table that these early quantum theorists were trying to understand. Figure 3 shows that periodic table. Note that there is no row corresponding to "actinides."

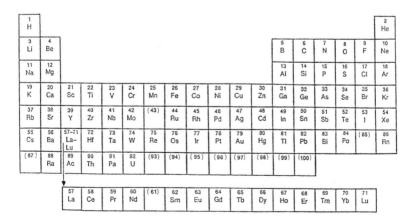

FIGURE 3 The pre–World War II periodic table.

In the modern periodic table the actinide row is made up of elements such as neptunium and plutonium, to say nothing of berkelium, which no one at that time had dreamed of. But there are basic commonalities. Take the family whose lightest element is helium. This family is missing in Mendeleev but was certainly known to these quantum theorists. Indeed, it had a suggestive and puzzling property. Helium has two external electrons. It is a "noble gas" in that it does not react with much of anything. The next element down is neon, which has added 8 electrons to make a total of 10, and then there is argon, which has added another 8 to make a total of 18. These too are noble gases with the same general chemistry. But why add 8? Why not 7 or 9? The very distinguished German physicist Arnold Sommerfeld, one of the masters of the old quantum theory, even suggested that it might have something to do with the number of vertices of a cube. This charming suggestion reminds me of Kepler's attempt to explain the planetary orbits by fitting them into the five perfect Euclidean solids (Plate 1).

In 1925, the outline of a solution to this puzzle emerged. It began in 1924, when the aforementioned Wolfgang Pauli, who was a student of Sommerfeld, proposed that the electron, in addition to its familiar properties, such as energy and angular momentum, had another quantum mechanical property that could take on one of two values—up or down, on or off—just two values. Pauli had no physical model for this property; he was trying somehow to account for the "magic numbers" in the periodic table. The following year two very young Dutch physicists from the university in Leiden, Samuel Goudsmit and George Uhlenbeck, neither of whom had their Ph.D.s, proposed a model. Goudsmit understood the details of the spectra, and Uhlenbeck recognized that they could be explained if the electron was assigned an additional angular momentum besides the angular momentum due to its orbital motion. The picture that they formed was the electron as a tiny spinning ball of electrical charge that, even when it was at rest, kept spinning. But this "spin" differed from that of, say, an ordinary top. Such a top can spin with its axis of rotation pointing in any direction. But this "top" could spin with the axis pointing in only one of two directions—up or down. This was their model for Pauli's two-valued quantum number.

They wrote a brief paper but then had second thoughts. They went to their professor, Paul Ehrenfest, and asked him to withdraw it, only to discover that he had already sent it to a journal. Soon the news was out and the reactions—mostly negative—set in. Pauli, remarkably, was against the idea. Indeed, at this very time Bohr was on his way from Copenhagen to Leiden where he was going to join Einstein and Ehrenfest. The train stopped in Hamburg and Pauli boarded it while it was in the station to warn Bohr about the spin. By the time the train reached Leiden, Bohr was quite concerned. He was met at the station by Einstein and Ehrenfest, and Einstein told him not to worry. It would work out. One of the objections had been put forward by the great Dutch physicist H. A. Lorentz. He argued that if he gave the spinning ball a plausible radius it would spin at its surface faster than the speed of light, which was forbidden by the relativity theory. Einstein must have understood that such classical considerations were not applicable to what was a clearly quantum theory property. Ironically, not long after, Pauli produced the mathematical formulation of the spin that we still use today.

That same year, 1925, Pauli proposed a principle without any justification except that it seemed to work, which cracked open the problem of the periodic table. This was what came to be called the Pauli "exclusion principle." He stated that two electrons (at first he applied it only to electrons and then later to other particles such as neutrons and protons that had the same spin as the electron) could never be in identical quantum states. To take a relevant example, if two electrons have the same energy and zero orbital angular momentum, then their spins, Pauli insisted, must point in opposite directions. If a third electron is added to the mix, it must have a different orbital angular momentum or energy, or both, because its spin would necessarily point in the direction of one of the two original electrons, which is not allowed unless it differs in some other respect.

We can see from this how the periodic table is going to be built up. The lightest element is hydrogen, and its electron can have its spin pointing up or down. Next is helium. Its ground state has no orbital angular momentum. The two electrons in it must have their spins pointing in opposite directions to obey Pauli. These two electrons form what is called a closed shell. Because they are paired, they are not available to participate in chemical reactions, which explains why helium is reluctant to form chemical combinations with anything.

The next electron must go into a new shell. The least energetic state into which it can go is one where it has no orbital angular momentum, but a somewhat higher energy than the zero orbital angular momentum ground state of helium. There is an old spectroscopic notation for these states. States of zero angular momentum are called "s states." The ground state is the 1s state, and the next most energetic state is the 2s state. The newly added electron is in the 2s state and outside the two-electron 1s state shell. We expect the corresponding element, which is lithium, to be chemically very active since this 2s electron can take part in chemical reactions. It is. To complete the 2s shell and obey the Pauli principle, just one more electron can be added. This corresponds to the element beryllium.

The next electron that can be added has, in Bohr's units, an angular momentum of 1. Unit angular momentum states are called p states. It turns out that there are only three different orientations that the quantum mechanical orbital angular momentum of one unit can have. Each of these electrons can have spin up or down so there are a total of six possibilities. Hence the shell with the s and p electrons closes when there are eight electrons. This is the element neon, which again is a noble gas. So we begin to see how the "magic numbers" are accounted for. To describe the rest of the elements one must go much more deeply into the theory.

I recall in the early 1950s, when I was looking for a thesis topic, Victor Weisskopf, who had been one of Pauli's assistants and was a group leader at Los Alamos, told me that there were still some unsolved problems with the periodic table. Somehow that seemed too "classical" for me, so I never found out what they were. I did learn, which I explain later, that the actinide series—the one that includes uranium and plutonium—is so strange that when Glenn Seaborg proposed an explanation for it in the early 1940s he was told that it was crazy and, if he insisted, he might damage his scientific reputation. Seaborg felt that, at the time, he did not have a scientific reputation to lose, so he persisted. With this excursion we can now return to the history of uranium and then plutonium.

IV
Frau Röntgen's Hand

At least until the end of the nineteenth century, rather little was done on the science of uranium. In 1841, the French chemist Eugène Peligot made it into a metal as dense as gold. It was also learned earlier that uranium salts and oxides could be used to produce wonderfully colored ceramics. We noted that Mendeleev included it in his 1872 version of the periodic table as the heaviest element. In 1895, however, a discovery occurred that was to change everything in the next few years. Wilhelm Conrad Röntgen was a professor of physics at the University of Würzburg in Germany. He had been experimenting with what were known as Crookes tubes. Traditionally these were glass cones from which the air had been pumped out. At one end there was a positively charged "anode" and at the other a "cathode," which emitted a stream of electrons that passed to the anode. The electron stream could be made to do various things such as causing minerals to fluoresce. Röntgen was curious as to how far the cathode rays—electrons—might penetrate outside the tube. He covered the tube with cardboard and turned on the electric discharge. What happened was more that he had bargained for. A barium platinocyanide-coated screen across the room began to glow. Curious, Röntgen held various materials between himself and the screen and much to his astonishment saw the bones of his hand. He called the mysterious beam "x-rays" and persuaded his wife to have her hand x-rayed. The picture he took (Plate 2), which also shows her wedding ring, was sent to colleagues. It was very quickly understood that x-rays were a new medical diagnostic tool of immense importance. Indeed, in 1901, Röntgen received the first Nobel Prize in physics for his discovery.

The next step involved one of those examples of the way scientific information propagates.[1] Henri Poincaré, the mathematician who was interested in everything, received a copy of the paper that Röntgen had written on his discovery. He had a colleague named Henri Becquerel who held a chair at the Museum of Natural History that also had been held by his father and grandfather. He even worked in the same field—the study of how substances such as uranium phosphoresced. Poincaré suggested that Becquerel look to see if substances might be giving off some of the new x-rays as they phosphoresced. Becquerel tried several but had no success until he used uranyl potassium sulfate. He had noticed that it became phosphorescent when exposed to the Sun. This phosphorescence made a photographic plate, even when covered with black paper, become foggy. For a few days after his first attempt the weather was bad, so Becquerel did not try to do further experiments. He put the photographic plate, which had been unexposed, and the uranium in a drawer. He was astounded to discover when he opened the drawer that the photographic plate had become fogged over. Something was emanating from the uranium. But what? This is what he wrote:

> One hypothesis which presents itself to the mind naturally would be to suppose that these rays whose effects have a great similarity to the effects produced by the rays studied by M. [Philipp] Lenard [who studied cathode rays-electron currents] and M. Röntgen, are invisible rays emitted by these bodies. However, the present experiments without being contrary to this hypothesis, do not warrant this conclusion. . . .[2]

By the time Becquerel delivered his Nobel Prize address in December of 1903 (he shared the prize in physics with Marie and Pierre Curie), the nature of the radiation had been at least partially clarified. In particular, four elements—uranium, thorium, radium, and polonium (the last two discovered and named by the Curies)—were known to be "radioactive," a term Madame Curie invented.

These elements gave off, variously, three types of radiation to which Rutherford had given the names "alpha," "beta," and "gamma," a terminology that Becquerel adopted. It was determined, by bending them in electric and magnetic fields, that the alpha and beta rays were electrically charged. The beta rays had one unit of negative charge and were soon identified as electrons. The alpha rays were trickier since, at first, they did not seem to bend in magnetic fields unless the fields were made much larger. It turned out that the alpha particles had masses some 8,000 times that of the electron. Also, they carried a positive charge. By the time of Becquerel's lecture, Rutherford and his colleague Frederick Soddy, who were both at McGill University in Montreal, had shown that the alpha particles, when collected in a gas, gave off the light spectrum of helium. Once the nucleus was discovered it became clear that the alpha particles were helium nuclei with two positive charges. Like the x-rays to which they were compared, gamma rays exhibited no electric charge. They are, indeed, a more energetic form of x-ray—an electromagnetic quantum. Becquerel noted that uranium emitted both beta and gamma rays, while thorium and radium seemed to emit all three.

The question that troubled Becquerel and everyone else was, What was the source of this seemingly limitless supply of energy? In his Nobel lecture he stated:

> Among the hypotheses which suggest themselves to fill the gaps left by current experiments, one of the most likely lies in supposing that the emission of energy is the result of a slow modification of the atoms of the radioactive substances. Such a modification, which the methods at our disposal are unable to bring about [no change in the environment

in which the radioactive elements were placed changed the nature of the radioactivity], could certainly release energy in sufficiently large quantities to produce the observed effects, without the changes in matter being large enough to be detectable by our methods of investigation.

It is not clear what Becquerel had in mind, but it would not be for another two years, until in the last of his great 1905 papers, that Einstein gave the correct answer. The residues of the decay—the "daughters" to employ the term of art—have less mass than the parent object that decays. This mass difference supplies the needed energy through the relation $E = mc^2$. Each time a particle decays it loses some of its mass. At the end of his 1905 paper, Einstein wrote, "It is not impossible that with bodies whose energy-content is variable to a high degree [because they decay] (e.g., with radium salts) the theory may be successfully put to the test."[3]

V
Close Calls

Prior to the 1930s, it must have occurred to people to ask if there were elements that had a higher atomic number than uranium. None had been found in nature. None ever were. There was a dramatic advance after 1932, when Chadwick discovered the neutron. Since this particle was electrically neutral, it was evident to many people that it would be the ideal probe of the nucleus. The physicist I. I. Rabi put the matter dramatically in 1934:

> Since the neutron carries no charge, there is no strong electrical repulsion to prevent its entry into the nucleus. In fact, the forces of attraction which hold nuclei together may actually pull the neutron into the nucleus. When a neutron enters a nucleus, the effects are about as catastrophic as if the moon struck the earth. The nucleus is violently shaken up by the blow, especially if the collision results in the capture of the neutron. A large increase in energy occurs and must be dissipated, and this may happen in a variety of ways, all of them interesting.[1]

The first people to suggest using neutrons in this way were the Joliots. In 1926, Frédéric Joliot married Marie Curie's daughter Irène, with whom he collaborated. They shared the 1935 Nobel Prize in chemistry. (The Nobel Prize in physics that year was won

23

by Chadwick.) However, they actually used alpha particles in their experiments. In their Nobel lectures the Joliots described their discoveries. Irène began her lecture by noting that Rutherford had been the first person to transform elements—that is, to perform modern alchemy. He bombarded elements such as oxygen and aluminum with alpha particles. A proton emerged—meaning, for example, that an aluminum atom had been transformed into an isotope of silicon. The Joliots were interested in "artificial" radioactivity. They bombarded elements that were not radioactive and created isotopes of elements that were. One of the remarkable things about their work was the miniscule amount of material they dealt with: samples of the order of a millionth of a billionth of a gram. Joliot noted that these samples comprised only a few million atoms, sufficient to give off enough radioactivity to be detected.

The hard problem was to determine chemically what these radioactive elements were. One of the techniques used had been exploited by Madame Curie. The particular example I am going to give is very important in our discussion of the discovery of fission. The problem she had was that her samples of radium were contaminated with barium. To remove the barium, she used what chemists call "fractional crystallization." This process takes advantage of the fact that, in a suitable solvent, different chemical compounds form crystals at different rates. Compounds have different degrees of solubility. The experimenter tries to concentrate what he or she is looking for in the crystals and remove the liquid. Madame Curie discovered that the radium salt was less soluble than the barium salt, so it became concentrated in the crystals. This was an immensely tedious task, but it worked. It enabled her to identify radium. The Joliots made use of the same techniques and were able to isolate the newly discovered radioactive isotopes, for which work they won the Nobel Prize. In his Nobel lecture Joliot offers some cautionary and rather prophetic remarks. Keep in mind that they were written in 1935.

If such transmutations do succeed in spreading in matter, the enormous liberation of useable energy can be imagined. But, unfortunately, if the contagion spreads to all the elements of our planet, the consequences of unloosing such a cataclysm can only be viewed with apprehension. Astronomers sometimes observe that a star of medium magnitude increases suddenly in size; a star invisible to the naked eye may become very brilliant and visible without any telescope—a Nova has just appeared. The sudden flaring up of the star is perhaps due to transmutations of an explosive character like those which our wandering imagination is perceiving now, a process that the investigators will no doubt attempt to realize while taking, we hope, the necessary precautions.[2]

What no one's "wandering imagination" could predict when Joliot gave this address was that four years later fission would be discovered.

Enrico Fermi and his group in Rome were the first scientists actually to use neutrons to explore nuclei. Before I get to this seminal work, I want to make a detour. I am going to talk about the life and times of Ida Noddack. The reason for this soon becomes clear. Ida Noddack, née Tacke, was born in 1896, in a suburb of what is now Wesel, Germany. Her father was a lacquer producer, which may be what led to her interest in chemistry. She was one of the first women in Germany to get any kind of advanced degree in chemistry, a master's and then a Ph.D. in 1921. In 1926, she married the chemist Walter Noddack. By this time they had already done successful work together (Plate 3) at the Imperial Physico-Technical Research Center in Berlin, where she had taken a job in 1925 and at which Noddack was the head of the chemistry laboratory.

The Noddaks were interested in filling in some of the gaps in the periodic table at that time. In 1925, along with an x-ray specialist named Otto Berg, they identified an element with an atomic number of 75 and an approximate atomic mass of 186, which was one of the missing elements. It was silvery white and metallic. The Noddacks were, if nothing else, patriotic Germans, so they chose the Latin name *rhenus* (Rhine) for it. It is now known as rhenium and has the symbol Re. Their German patriotism went, many people would argue, somewhat too far. While it was never shown that they were Nazis, they certainly weren't anti-Nazis. In fact, in 1942, Walter Noddack accepted a position at the recently reopened university—now as a Reich university—in Strasbourg, which the Germans had taken from the French in 1940. One of their fellow academics in this now suitably aryanized institution was the aforementioned C. F. von Weizsäcker. Tenure was not long since the Allies retook Strasbourg in 1944. About the time the Noddacks found rhenium they also claimed to have found another missing element, 43, which Walter Noddack wanted to call "masurium" after his homeland in East Prussia. This one was disputed, and its discovery is usually attributed to two Italians (one of them being Emilio Segrè, who will play an important role in our story) in 1937. They called it technetium, its present name. If you look at the prewar periodic table, you will see that rhenium is in the column just above the first missing transuranic element (93). From what has been said earlier, you should have no problem understanding why, until its actual discovery, this element was often called "ekarhenium."

Now we can return to Fermi. Noddack comes back into our story in a most remarkable way. In his autobiography, *A Mind Always in Motion*,[3] Emilio Segrè notes that for the fun of it the young physicists around Fermi decided to give themselves ecclesiastical names. Segrè was known as the Prefect of Libraries because of his interest in physics literature. Orso Corbino, who ran the physics department and was also a state senator, was known as the Heavenly Father. Ettore Majorana, who was a brilliant and very critical theorist, was known as the Grand Inquisitor. (He disappeared under mysterious circumstances in March 1938 while taking the Palermo–Naples ferryboat.) Fermi was, of course, known as the Pope. Fermi was probably the last physicist who knew all of physics. He did fundamental work in every branch, both theoretical and experimental. The subject is too diverse now for anyone to have such mastery. At about the time he was doing the experimental work I am going to describe, he developed the first real theory of those radioactive processes in which electrons are produced: beta decay. We still use his basic ideas today. He also created the statistics for the class of particles to which electrons and neutrons belong: particles with up–down spin. In his honor, we call these particles "fermions." He wrote a classic pedagogical review article on quantum electrodynamics that instructed generations of physicists. But the Nobel Prize in physics that he won in 1938 was not awarded for any of these. He won it for his experimental work with neutrons, some of which, as it turned out, was wrong.

When Fermi learned about the discovery of the neutron he immediately began making preparations with his group to use it to penetrate nuclei. In 1934, his group heard about the work of the Joliots and began an intense series of experiments in which they irradiated every element they could obtain with neutrons. Segrè was put in charge of procurement and even managed to borrow a gold ingot from a Roman firm Staccioli. They went systematically through the elements until they reached uranium. By using uranium as a target, they were quite certain that they had produced the first transuranic. Fermi called them "transuranes." Indeed, Fermi wrote a brief paper, which he published in the British journal *Nature*,[4] entitled "Possible Production of Elements of Atomic Number Higher than 92," in which he explains that when elements are bombarded, three possible results are observed: capture of a neutron with the emission of an alpha, beta, or gamma ray. He notes that when uranium is bombarded, something is produced that emits beta rays and seemed to have a half-life—the time in which half of any sample decays—of 13 minutes. Fermi goes on to describe the tests that were done on this object and argues that they exclude the possibility that it can be any of the known elements heavier than lead and lighter than uranium. He then notes:

> The negative evidence about the 13-minute activity from a large number of heavy elements suggests the possibility that the atomic

number of the element may be greater than 92. If it were element 93, it would be chemically homologous with manganese and rhenium. This hypothesis is supported to some extent also by the observed fact that the 13-minute activity is carried down by a precipitate of rhenium sulphide, insoluble in hydrochloric acid. However, as several elements are easily precipitated in this form, this evidence cannot be considered as very strong.

Manganese and rhenium, as one can see from looking at the prewar periodic table, are in the same column as element 93, which is why they were thought to be "chemically homologous." It is interesting to note that between them was element 43—technetium, to use the modern term. This is the element that the Noddacks claimed to have identified. It is equally interesting that Fermi does not mention it, which leads one to wonder if he shared the widespread skepticism about this claim. If so, it may help explain the remarkable sequence of events that followed.

One of the readers of Fermi's paper was the aforementioned Ida Noddack. She did not believe his results and said so in no uncertain terms in a brief paper published shortly after Fermi's. Noddack begins her paper by remarking that earlier that year she herself had published a paper in which she had discussed the possibility of finding transuranic elements. She writes: "A few weeks later it was reported, first in the newspapers, and then also in technical journals, that two scientists, Professor Fermi in Rome and Mr. Koblic in Joachimsthal, independently had discovered the element with number 93." Odolen Koblic was a Czech engineer who thought he had found the transuranic by analyzing pitchblende, and named it "bohemium," but later withdrew the claim.[5] Fermi did not cite Noddack's paper.

She goes on to describe Fermi's experiment and then comes the criticism. She writes:

> Fermi was able to make a chemical separation of one of the new radio elements, which had a half life of 13 minutes. He did this by adding manganese salt and concentrated nitric acid to the ura-

nium nitrate, then heating to the boiling point and adding sodium chloride. The resulting manganese dioxide precipitate was found to contain almost all the beta activity with the 13-minute half life [the alleged transuranic]. Fermi next tried to show that the radioelement which is responsible for this beta activity was not an isotope of any known element near uranium. To show this he added known beta emitting isotopes of the following to the acid solution of uranium nitrate: protactinium (91), thorium (90), actinium (89), radium (88), bismuth (83), and lead (82). When sodium chloride is added to precipitate the manganese dioxide none of these beta-emitting isotopes are found in the precipitate, according to Fermi. Since the unidentified new radio element does precipitate with manganese, and since it could not be an isotope of radon (86) or francium (87) either according to its properties, Fermi concludes that it might be the unknown element 93 (or perhaps 94 or 95).

In short, Fermi rules out as the new element any known element between lead and uranium in mass. But now Noddack makes her point: Fermi stopped his analysis too soon. He should have ruled out not only these heavy elements but the rest of the periodic table as well. How does he know that it is not one of the other elements?

Then comes one of the most remarkable observations in all the physics papers of this period. It is remarkable both for what it says and for the fact that it played no role at all in subsequent events. Ida Noddack writes:

One could assume equally well [equally well to having created a trans-uranic] that when neutrons are used to produce nuclear disintegrations some distinctly new nuclear reactions take place which have not been observed previously with proton or alpha-particle bombardment of atomic nuclei [the kind that the Joliots and Rutherford were carrying out]. In the past one has found that transmutations of nuclei only take place with the emission of electrons, protons, or helium nuclei, so that the heavy elements change their mass only a small amount to produce neighboring elements. When heavy nuclei are bombarded by neutrons, it is conceivable that the nucleus breaks up into several large fragments, which would of course be isotopes of known elements but would not be neighbors of the irradiated element.[6]

Putting the matter simply, in September of 1934, Noddack was suggesting that what Fermi observed was nuclear *fission!* Some explanations as to why this was ignored are suggested after I describe a discovery that Fermi made at this time—perhaps from a practical point of view his most important discovery—which makes it even more bizarre that, with or without Noddack, the group in Rome did not discover fission.

In the fall of 1934, Segrè and the others began noticing a very odd effect. When they irradiated their elements with neutrons, the intensity of the effect seemed to depend on what sort of material the apparatus was sitting on. If it was sitting, for example, on wood, there was a sharp increase in the intensity. This seemed to make no sense at all and made their experiments difficult to interpret, so they decided to investigate. What happened next is another of those examples of "sleepwalking" that great scientists sometimes have. Recall that Rutherford asked his young protégés Geiger and Marsden to keep a lookout for alpha-particle collisions with the gold foil at which the alpha particle would emerge from such a collision at a wide angle. My guess is that if anyone had asked Rutherford at the time why he was making that suggestion he would have had no idea. Many years after the fact, Fermi described his sleepwalking to his Nobelist colleague at the University of Chicago, the astrophysicist Subrahmanyan Chandrasekhar. This is what he told Chandrasekhar:

> I will tell you how I came to make the discovery which I suppose is the most important one I have made. We were working very hard on the neutron-induced radioactivity and the results we were obtaining made no sense. One day, as I came to the laboratory, it occurred to me that I should examine the effect of placing a piece of lead before the incident neutrons. Instead of my usual custom, I took great pains to have the piece of lead precisely machined. I was clearly dissatisfied with something; I tried every excuse to postpone putting the piece

of lead in its place. When finally, with some reluctance, I was going to put it in place, I said to myself: 'No I don't want this piece of lead here; what I want is a piece of paraffin.' It was just like that with no advance warning, no conscious prior reasoning. I immediately took some odd piece of paraffin and placed it where the piece of lead was to have been.[7]

This took place in the morning of October 22nd. Much to their astonishment, putting the paraffin between the neutrons and the target element enhanced the nuclear reactions enormously. At first they thought that something had gone wrong with their counters, so they tried other substances as neutron filters and nothing much happened. It was only paraffin that gave rise to the dramatic effect. It was now lunchtime, and Fermi went home for lunch and a siesta. When he returned at three, he had created a new branch of physics. Paraffin is a hydrocarbon. It consists of hydrogen and carbon atoms. What Fermi realized was that fast neutrons colliding with the hydrogen or carbon nuclei would be slowed down. The hydrogen nucleus—the proton— would be, from one standpoint, the best at slowing neutrons down. It and the neutron have about the same mass, so that in a collision the proton could take away a good deal of the neutron's momentum, thereby slowing it down substantially. However, there is a problem. Protons can sometimes capture neutrons. A proton and a neutron can bind together, creating the nucleus of heavy hydrogen—the deuteron. Once this happens the neutron is no longer available for probing the target nuclei. So while hydrogen is useful, it is not ideal. On the other hand, carbon is. It does not capture neutrons but just slows them down. When, on December 2, 1942, Fermi demonstrated the first nuclear reactor, which was built on some squash courts under Stagg Field at the University of Chicago, the neutrons that were slowed down to enhance their reactivity were "moderated" using purified graphite, pure carbon. Fermi's paraffin filter was the prototype of a moderator in a reactor. To a lesser degree, the wood on the table had acted in the same way because it also contained hydrogen and carbon.

This explained what the neutron did. It slowed down. But why did this increase the reaction rates? That is more subtle. A neutron is not a simple projectile like a baseball. In some situations it does act like a classical particle with a mass and momentum. We do not expect classical particles to become more effective in, say, breaking a window, when they are moving more slowly. But a neutron, like all the "particles" in quantum theory, also has the characteristics of a wave. This changes the properties of the collisions. A measure of the effectiveness of one of these collisions is the effective area presented by the target, which is called the "cross section." If the neutron were a classical particle, this area would be the same for neutrons of any speed. But because of the wave nature of the neutron, the effective area can increase as the neutron is slowed. Indeed, for reactions of the kind that Fermi was studying, the cross sections increase as $1/v$, where v is the speed of the neutron. The basic reason is that the slower the neutron is moving, the more time it spends in the neighborhood of the nucleus with which it interacts.[8] This is the effect that the group in Rome had observed; with it the science of collisions with slow neutrons, one of the things for which Fermi was awarded the Nobel Prize, was born. But the collision of slow neutrons with uranium was just how fission was discovered four years later in Germany. The question is, Why did Fermi and his group miss it four years earlier?

Some years ago I had the chance to spend an evening with Segrè and asked him this very question. Later, when he wrote his biography of Fermi, he repeated what he had told me.[9] To understand his answer I need to explain a little more of the physics of fission. The next chapter, which deals with the discovery of fission, provides more details. As Noddack correctly pointed out, fission is a process in which a heavy nucleus such as uranium splits into lighter nuclei, presumably to be found somewhere in the middle of the periodic table. A great weakness of Noddack's paper is that she does not supply a specific example of such a process. She never says what nuclei might be involved. Fission, it turns out, can either occur spontaneously or be induced when a neutron is absorbed by the fissioning nucleus. In either case, energy must be conserved. This is a fundamental law of physics. It means that the masses of the final products must add up to a mass less than the mass you start with. What is conserved is the total energy, including the mass energy. The loss of mass goes into the kinetic energy of the fission products. This is what Einstein taught us. For spontaneous fission, this is the mass of the parent nucleus, while for the induced fission it is the mass of the parent plus the mass of the neutron that initiates the reaction. Without knowing the masses involved, we cannot say whether the energy released in the fission—which is proportional, according to Einstein's equation, to the difference in masses between the initial and final particles—is large or small. In fact on a nuclear scale it turns out to be quite large.

Not knowing the masses, neither the Rome group nor Noddack would have had any way of determining the mass loss. But in the experiments that the Rome group was doing, fission was certainly taking place. Energy pulses were being produced. Why didn't they see them? This is what Segrè explained. At the time they were focused on transuranics. They thought that a transuranic might give off an energetic alpha particle more rapidly than their timing detectors could register. Indeed, they were not detecting any alpha particles. They decided that if they covered their sample of uranium with aluminum foil, only the energetic alpha particles would get through and be observed. But what they did not realize was that the shield was thick enough to block the pulses from the fission. They never saw them. As Segrè writes, "It was this aluminum layer that prevented us from seeing the big ionization pulses characteristic of fission, but it is impossible to say whether we would have correctly interpreted the phenomenon if we had observed it."[10] Given Fermi's genius it is difficult to imagine that he would not have interpreted the result correctly.

I also asked Segrè why they had ignored Noddack's paper. He was not sure. Several explanations have been offered by other people. Some people have argued that it was because she was a woman.

Possibly, but I don't think so. If, for example, Irène Curie had written the paper, I am sure they would have paid attention to it. The Noddacks had a mixed reputation. While they did discover rhenium, their claim to element 43, which they vigorously defended, was considered by the scientific community to be very dubious. There is also the fact, as I have previously mentioned, that Ida Noddack does not give a specific example. She does not consider the energy question at all. There is also the tone of her paper. Her criticism of Fermi involved not only his science but the fact that his discovery of transuranics, which turned out to be incorrect, was announced in the newspapers. I do not think that Fermi was someone who took kindly to this sort of criticism. More interesting questions are: What would it have meant if Fermi and his group had discovered fission in 1934, four years before its actual discovery? Would the nuclear arms race have started in 1934? Who would have participated in it? Would the Second World War been nuclear from the beginning? We will never know.

VI
Fissions

*The Hahn discovery was checked in many labo-
ratories, particularly in the United States, shortly
after publication. Various research workers (Meitner
and Frisch were probably the first) pointed out the
enormous energies which were released by the fission
of uranium. On the other hand, Meitner had left
Berlin six months before the discovery and was not
concerned herself in the discovery.*

Document signed by the 10 Farm Hall detainees,
including Otto Hahn, on August 8,1945[1]

Lise Meitner (Plate 4) was born into a comfort-
able middle-class Jewish family in Vienna in 1878.[2] Her father was
a lawyer who was able to provide well for his eight children of which
Meitner was the third. He believed that young women should have
the same educational opportunities as their male counterparts, but
formal schooling for women in Vienna stopped at age 14. They were
then supposed to return home and prepare for marriage. The only
educational opportunities were provided by private tutors, and with
a few other young women, Meitner studied physics and mathematics
with tutors. This was in preparation for an examination known as the

Matura, which would decide whether or not she would be accepted to the university. She passed and enrolled in the University of Vienna in 1901. She was already 23.

Meitner had the good fortune of being able to take courses with one of the greatest physicists of that era—Ludwig Boltzmann. Not only was Boltzmann a deep and original physicist, but he was also, it seems, a wonderful teacher—a rare combination. Meitner often described the profound impression that Boltzmann's lectures made on her. She was not alone. My teacher Philipp Frank, who was a few years younger, was at the university at about the same time. He also took courses from Boltzmann and even did his Ph.D. with him. Professor Frank told me that of all the lecturers in physics, from Einstein on down, whom he had heard, Boltzmann was the best. But Boltzmann was a depressive. He had tried to commit suicide more than once before he succeeded in doing so in September of 1906. By this time, Meitner had already gotten her Ph.D., working on an experimental problem. She was the second woman in Austria to get a Ph.D. in physics. Throughout her career, Meitner was an experimental physicist with strong attachments to theory, although, unlike Fermi, I am not aware of any important contributions she made to theory, fission excepted. After Boltzmann's death, Meitner remained in Vienna for a year continuing to do physics research and teaching school. But she realized that if she stayed, she would end up as a school teacher, so with the financial assistance of her parents, she went to Berlin to continue her studies. On September 28, 1907, she met Otto Hahn for the first time.

Otto Hahn (Plate 4) was four months younger than Meitner. He had been born in Frankfurt and was the youngest son of a prosperous glazier. He had studied chemistry at the University of Marburg, where he was known as a good, but not particularly dedicated, student. He had gotten his Ph.D. in 1901 and became a lecturer at the university. In 1904, he decided to go to England to learn English as well as more chemistry. There is perhaps some irony in this because, in 1945, he ended up as one of the Farm Hall detainees, several of whom spoke

English. While in England as a young Ph.D., Hahn switched his field to radiochemistry and soon discovered a new radioactive isotope of thorium. At the time, the notion of isotopes was not yet known, so Hahn thought he had discovered a new element. After a year in England he went to Montreal to work with Rutherford. In 1906, he got a junior position in the Berlin laboratory of the very distinguished organic chemist Emile Fischer, who had won the second Nobel Prize in chemistry in 1902. In Berlin, Hahn was able to pursue his interest in radioactivity. Fischer had no idea of, and no interest in, what Hahn was doing with radioactivity, but paid him a small salary. Hahn had no teaching responsibilities. Meitner, on the other hand, although she was offered a nominal unpaid position in the laboratory of another senior professor, Heinrich Rubens, had to find work space in Fischer's Chemistry Institute. Fischer wanted nothing to do with women scientists. So Meitner was relegated to a small space in the basement in a building that had no lady's room. She used the facilities in a nearby restaurant and was told that she could not enter student laboratories where there were male students.

What the relation between Hahn and Meitner was is not clear, at least to me. Ruth Sime in her biography[3] of Meitner paints a very charming picture. She writes of Hahn's "good-natured informality"[4] and explains how they liked each other's company as an escape from the rigid formality of German academic life. On the other hand, in his review of Sime's book,[5] Max Perutz writes:

> In their relations with each other, Meitner and Hahn never deviated from the strict Victorian code for relations between the sexes: they addressed each other as Fräulein Meitner and Herr Hahn, and avoided eating or going out for a walk together, signs of intimacy that might have invited gossip. It took sixteen years and the post–First World War revolution before they called each other Lise and Otto and used the familiar *du*.

This seems more nearly correct to me, and later in her book Sime corroborates this. Here is some evidence of Hahn's formality dating from a much later time.

Erich Bagge was a young German physicist who was interned in Farm Hall with the nine others. The British made capsule descriptions of each detainee that included their photographs. Later I present their description of Hahn, but here is what they said about Bagge: "A serious and very hard-working young man. He is completely German and is unlikely to cooperate."[6] Indeed, the Bagge of the transcripts is not a very likeable individual, but he kept a journal. In it he reveals himself to be a much more sympathetic figure. The journal gives another view of Farm Hall apart from the transcripts. Here is Bagge's entry for Sunday, November 18, 1945:

> Today I can report an interesting event, one without peer. Friday morning, shortly after breakfast, most of us were sitting in the drawing room in order to listen to "this week's composer"—it was Rimsky-Korsakov—and naturally to study the "latest news" [the Germans were provided with newspapers to see if they would reveal anything by their reactions] when Heisenberg said to Hahn: "Mr. Hahn, take a look at this!" Therewith he handed him the *Daily Telegraph*. Mr. Hahn, who was then himself busily reading another paper, said: "I don't have the time." "But it is very important for you; it says that you are supposed to receive the Nobel Prize for 1944." The excitement that struck the ten detainees is hard to describe in a few words. But gradually we broke through with Heisenberg in the lead, who congratulated him heartily on the 6200 pounds [the value of the Nobel Prize at this time]. Then the rest of us succeeded in turn. Heisenberg immediately went to the Captain [the British officer in charge], who was completely surprised by the news, and was still totally stunned a half hour later. He immediately called the London Office. Apparently nothing was known there as yet, but the telephone was buzzing. The Information Ministry, *Times* correspondents, and all sorts of newspaper people were called. It gradually became apparent that we were not dealing with a false report.[7]

At the end of this chapter I return to the discussion of Hahn's Nobel Prize. One is of course struck by the fact that he was given it for the discovery of fission three months after Hiroshima. But I quoted the passage as an illustration of the formality in German academic life. At this point in Bagge's journal, the detainees have been living together in a moderate-size country house for more than five months. They have taken all their meals together and socialized constantly. But, as the excerpt shows, even so, Hahn and Heisenberg refer to each other as Mister Hahn and Mister Heisenberg. One is therefore not surprised at the formality that existed for many years between Meitner and Hahn. Nonetheless, they made an excellent collaborative team. He had not studied mathematics and physics at the university, and she was not trained as a chemist. He was a superb observational chemist, and she supplied the theory, including the plotting of graphs.

They had two periods of collaboration: from the time of their meeting until the end of the First World War and then from 1934 until just before the discovery of fission in December 1938. In between, Meitner and Hahn worked on separate problems with their own students and assistants. She began a long experimental study of beta decay, something she had started with Hahn. She had a then-common, but mistaken, idea that she defended doggedly until experiments done by others at the end of the 1920s finally forced her to change her mind. This idea had to do with the energy with which the electron was emitted in beta decay. Suppose you have a collection of identical beta-decaying atoms each one emitting its electron. The natural assumption that Meitner, and nearly everyone else, made at the time was that all of these electrons would be emitted with the same energy. This assumption turns out to be erroneous, but the experiments that showed it were very difficult to conduct and to interpret. Each of the identical atoms emits an electron with a different energy, so that the ensemble of atoms produces electrons with a spectrum of energies. This phenomenon seemed to violate the conservation of energy and momentum, but in 1930, Pauli suggested that in addition to the electron, a light, electrically neutral, and hence undetectable particle was also emitted that carried off some of the energy and momentum. Fermi adopted this idea in his theory of beta decay, which predicted the energy spectrum of the electrons. He named the particle the "neutrino"—little neutral one—to distinguish it from the "neutrone"—big neutral one—the neutron.

The most significant problem that Hahn and Meitner tackled in their first period of collaboration was the following. Uranium is radioactive, and it was then well known that a sequence of decays that began with uranium eventually produced the element actinium, a metallic element with the atomic number 89. It was also known that the last step of this series involved an element that had a long-lived alpha decay, in which a helium nucleus was emitted. From this property, one could deduce a good deal about the actinium precursor. Since an alpha particle has two positive charges—two protons—in its nucleus, it followed that the precursor had two more positive charges in its nucleus than actinium; that is, its atomic number must be 91. The pre–World War II periodic table, which we have seen, shows that this element was predicted to be in the same row as tantalum, another metal, with atomic number 73. Naturally, since it was one element beyond tantalum in the row, it was referred to as "ekatantalum." It was assumed to have the same chemical properties as tantalum. This assumption turned out to be wrong, however, which complicated the analysis. A short-lived isotope of ekatantalum had been identified in 1913 by a Polish chemist named Kasimir Fajans working with a German chemist named Oswald Göhring in Karlsruhe. There was some controversy about this discovery, so Hahn and Meitner decided to study the question. They started this work in 1913 and found a long-lived isotope. By the time they wrote their paper in 1918, a great deal had happened in their personal and professional lives.

It had been recognized for several years that German chemistry needed a national laboratory devoted to research that would attract the young chemists, many of whom were emigrating. Thus, in 1912 the Kaiser Wilhelm Institute for Chemistry, located in Dahlem, a suburb of Berlin, was opened, along with other Kaiser Wilhelm scientific institutes, all of which were independent of any university, so their scientists had no teaching obligations. Hahn was offered a junior position that carried the title of "Professor" and paid a good salary. Meitner was only offered the position of "guest," which paid no salary. But by this time, she was well known in the Berlin scientific community and had developed a friendship with Max Planck, who was probably the most important physicist in Germany. He was also a very decent man. He must have recognized that Meitner was being treated poorly because he made her one of his assistants that year. This was the first scientific job Meitner had that actually paid a salary. In 1913, she received an appointment on exactly the same level as Hahn's, but with a lesser salary, at the institute. They now had a joint laboratory in both their names. A year later, the First World War began.

At the outset, both Meitner and Hahn were very enthusiastic about the war. She returned to Vienna to see her brothers off to the front; but when she returned to Berlin she was subject to some of the Socratic dialogue that Einstein was carrying out, arguing that the war was idiotic. She thought at the outset that he was naïve but during the course of the war she began to understand the costs. She learned to operate x-ray machines, which she used in hospitals close to the front. She saw wounded and dying soldiers who affected her deeply. Remarkably, on the other side of the lines, Madame Curie and her daughter Irène were doing the same thing for French soldiers. Like many German men, Hahn was in the army reserve and was called up for active duty. He saw hard action in places such as Belgium. In January 1915, he left the front lines for a new activity: making poison gas. Whether he volunteered for this or was ordered to do it is difficult to say. But he was fully aware of the effect of this gas on enemy troops. Hahn was not a man given to much moral introspection. As far as I know, he never stated that the use of poison gas was morally wrong. He said that if he hadn't done it, someone else would have or that he did it because he thought it would shorten the war. One wonders what he could have meant by this extraordinary statement. In his autobiography the only thing he has to say about it is:

> Early in 1915 I became a lieutenant in the *Landwehr* and Professor Fritz Haber [a Nobel Prize–winning chemist who led the poison gas effort] saw to it that I was transferred to a group of "active specialists" [in what?]. In 1917 I was transferred to Supreme Headquarters, and in this capacity had official contact with the military research [On what? He does not tell us it was poison gas] carried out in Haber's Institute in Dahlem.[8]

In January of 1917, Meitner was invited to create a physics department within the chemistry institute. She now had her own laboratory which ran in parallel to Hahn's. Since Hahn was still involved in the war effort, it fell to Meitner to do most of the work of finding the missing isotope. She also wrote the paper. For whatever reason, she named Hahn as the senior author. Hahn seemed to find this acceptable. He also found it acceptable to receive a medal from the Association of German Chemistry for, among other things, the identification of the isotope of the element ekatantalum, which they had named "protoactinium" (later shortened to protactinium). The previous claim to the discovery was discounted. Meitner was given a copy of the medal. She certainly never complained publicly about this and even wrote Hahn a congratulatory note. But while they remained friends and colleagues, she did not collaborate with him again for more than a decade.

By the time their collaboration resumed in 1934, the political situation in Germany had changed radically: The Nazis had come to power. This event affected Hahn and Meitner in different ways. In 1908, Meitner had been baptized. She had never had any connection with the Jewish community, and two of her sisters had converted to Catholicism. Very likely, Meitner wanted to harmonize with the German community and saw her Jewishness as an unnecessary piece of baggage. It didn't do her much good. In 1923, she had been made a professor at the University of Berlin. In 1933, the racial laws were enacted that forbade anyone with any Jewish grandparents to teach in a university, so she was fired. On the other hand, she was an Austrian citizen and, so long as Austria was an independent country, she was allowed to keep her position at the institute, since it had been set up to be independent of any university. With Hahn it was a different matter. He was never a Nazi Party member, nor did he have Nazi sympathies. But like many, indeed most, of his German fellow scientists he was willing to make whatever compromises were necessary to keep his life as comfortable as possible and to preserve his institute, even if it meant firing people of greater competence and replacing them with Nazis, who in the end ran the institute. In view of this, the hiring of Fritz Strassmann (Plate 5) was remarkable. In many ways, Strassmann is the unsung hero of the fission story. Strassmann was a 32-year-old, very gifted chemist who, in 1934, joined Hahn and Meitner at the institute. He was an outspoken anti-Nazi. He had refused to join a Nazi teachers' union, which disqualified him for any university job. In fact, he and his wife were nearly starving. Despite this depredation, they hid a Jew in their apartment for a while. If they had been found out, everyone concerned would have been executed. Hahn and Meitner were able to pay him enough to see him through, but by the end of the war he was emaciated. Under these conditions, the three of them began a collaboration in 1934, which four years later ended with the discovery of fission.

After she heard of the discovery of the neutron, Meitner began her own experiments. It is instructive to ask where the neutrons came from. This was before the invention of the cyclotron, which became the neutron source of choice. Meitner, Fermi, and the other neutron experimenters of the time used natural radioactive sources such as radon—element 86—which, when it decays, produces an alpha particle with a good deal of kinetic energy. If this alpha particle is made to impinge on a beryllium target, neutrons of fairly high energy are produced. Meitner was also able to generate neutrons of lower energies by using the gamma radiation emitted by radium to interact with the beryllium. She noticed that for the neutron reactions she was studying,[9] the rates were increased for the lower energies. After she saw Fermi's papers, she was able to write him about her observations that confirmed his. She also studied his claim to have found ekarhenium—element 93—and decided to pin this elusive element down once and for all. She and Hahn began this study together and were soon joined by Strassmann. Throughout their joint work they adopted the canonical mantra:

A. Elements produced when neutrons impinge on a target can only have masses a few atomic masses different from the mass of the target.

B. Transuranics are "eka" elements that would be chemical homologues to the corresponding nontransuranics. For example, ekarhenium would be a chemical homologue of rhenium.

Both of these propositions turned out to be false.

Their plan of attack was the following. They did not think that because of the weakness of their radiation source, they would have enough neutrons to generate more than a minuscule amount of the new elements or isotopes, probably too little to do a decent chemical analysis. There would be enough, however, to detect whatever decay products—electrons, for example—might be emitted by the radioactive detritus produced by the neutrons. This radiation could be detected in a Geiger counter, for example. They could plot curves that showed how the decay rates—the number of electrons, say, produced per second—varied with time. From such a plot, one can in principle read off the half-life of the decay. Things become complicated if more than one isotope is decaying, but this is the general idea.

By 1936, they had irradiated uranium with both fast and slow neutrons and had observed some 10 radioactive species produced by the neutron collisions. Between them, Hahn and Meitner knew as much about such decays as anyone, and they concluded that some of the ones they were observing did not correspond to any known element or isotope. Because of their mantra, they were persuaded that they were observing transuranics. In fact, they claimed to have discovered four of them: elements 93 through 96. We have discussed 93 (ekarhenium), and 94 (ekaosmium) is the subject of this book. Its real discovery is the subject of the next chapter. Elements 95 and 96, ekairidium and ekaplatinum, respectively, were only discovered in 1944 and then named americium and curium, respectively. The 1936 Hahn–Meitner–Strassmann paper exuded confidence about the discovery of the transuranics. There was a voice of dissent—naturally, that of Ida Noddack—and needless to say no one paid any attention to her. The greatest irony of their work was that some of the activities they were observing actually came from radioactive fission fragments that they refused to believe were there. A close-second irony is that they observed a 23-minute activity that they thought might be coming from a uranium isotope, but they did not investigate further. In fact, it was coming from uranium-239, an isotope that decays too rapidly to be found in nature. What it decays into is just element 93, ekarhenium. Thus, they missed the real discovery of the first transuranic. The last paper the three collaborators published together on neutron interactions, or indeed on anything else, was in 1938, and by that time the roof had fallen on Meitner.

Even though the Kaiser Wilhelm Institute for Chemistry, where she had her laboratory, was now being run by people with Nazi affiliations, Meitner was still protected by her Austrian citizenship. But in March of 1938, the German army marched into Austria and it became part of Germany. Immediately Meitner was in danger.[10] She was denounced as a Jew. Hahn reacted by going to see one Professor Heinrich Hörlein, who was the treasurer of the Emil-Fischer-Gesellschaft, which was the actual sponsor of the institute, to grant an exemption to Lise Meitner. Hörlein demanded that Meitner leave the institute, and as a result of this conversation, Hahn told (more or less ordered) her not to come back. She felt betrayed. Hahn was putting his self-interest above that of a friend and collaborator of some three decades. But even apart from this betrayal, there was a question of attitude. He was treating her like some sort of assistant who could be hired and fired by him instead of as the head of her own laboratory. There is some suggestion that this is the way he really felt. As I mentioned, in the Farm Hall transcripts there is a brief section in which the photographs of the detainees are shown along with a capsule description written by their British overseers.

Under Hahn's photograph is written:

> The most friendly of the detained professors. Has a very keen sense of humor and is full of common sense. He is definitely disposed to England and America. He has been very shattered by the announcement of the use of the atomic bomb as he feels responsible for the lives of so many people in view of his original discovery. He has taken the fact that Professor Meitner has been credited by the press with the original discovery very well, although he points out that she was in fact one of his assistants and had already left Berlin at the time of his discovery.[11]

One notes in this absurd characterization no mention of Strassmann, whose role was absolutely essential. Meitner continued to work, hoping that somehow she would be made an exception to the disaster that was befalling her fellow Jews.

People living abroad seemed to recognize the danger more clearly than did Meitner. She had had the opportunity to emigrate to the United States but turned it down because she felt that, approaching her sixties, it would be too difficult to adjust. By May of 1938, she realized that she was in real danger and had to flee for her life. She requested permission to leave Germany, but this was not forthcoming. By mid-June she realized that such permission would not be granted to her. She would not be issued a passport. But Meitner had colleagues in Holland—above all, the physicist Dirk Coster—who were determined to save her life. In mid-July, Coster came to Berlin to escort Meitner to Holland personally. On the 13th she boarded a train with him. At the station to see her off was Hahn. He gave her a diamond ring, which he had inherited from his mother, to sell in case of an emergency. I have never been able to find out if she sold the ring or gave it back to him after the war. Apart from the ring, she had no money at all. She simply had two small suitcases with her clothes. She and Coster were able to cross the border, and Meitner found herself in Holland jobless and stateless. She knew that she could not stay in Holland. There was no job for her and above all she wanted to continue to work. It was fortunate that she didn't stay, given what ultimately happened to the Jews there. She was offered a job in Sweden, which she accepted, arriving in Stockholm on August 1, having flown first to Copenhagen, where she visited both Bohr and her nephew Otto Frisch, a Viennese-born physicist who had found refuge in Bohr's institute. No doubt, the theory of fission would have been discovered by someone, but the fact that it was found in the winter of 1939 by Frisch and Meitner in Sweden is one of those historical accidents that are truly unpredictable.

Back in Berlin, Hahn and Strassmann continued with their work. Hahn and Meitner were able to exchange letters that had both a scientific and a personal content. She wanted to know the latest news from the institute, in particular who was going to replace her, and she wanted her belongings sent to her, which turned out to be very difficult. They were able to meet once in Copenhagen, something that Hahn kept secret. He and Strassmann had hit a stone wall, which is one of the things he discussed with Meitner. Using slow neutrons, which enhanced the reaction rate, impinging on uranium, they had produced what they thought were radium isotopes. But uranium is element 92 and radium is 88, which meant that two alpha particles had to be produced in this reaction since there was a difference of four positive charges between the nuclei of these elements, and each alpha particle had two positive charges in its nucleus. The idea that a slow neutron could produce two alpha particles seemed totally crazy. Therefore they decided to study more closely the "radium" that had allegedly been produced. To do this they used variants of the fractional crystallization method, which had been invented by Madame Curie. The general idea was to put the so-called radium in some sort of solution to which a barium salt (a compound of barium) was added. They expected that crystals containing the barium salt and the "radium" would form. When they drained the remaining solution, what would be left over would be crystals with this radium, more highly concentrated than it had been before. Strassmann prepared a witches' brew consisting of barium chloride and the "radium" in a solution of nitric acid. Beautiful crystals formed, but there was no change in the radioactivity whatsoever. They were totally baffled. They tested the method by using real radium and everything went as it should have done. The only conclusion that Hahn could reach was that their "radium" *was* barium. Somehow the uranium nucleus had split into pieces, one of which was barium, element 56 in the periodic table—36 units away from uranium—with an atomic mass of 137 compared to the 238 of uranium. They were dumbfounded. There was no explanation for this result whatsoever.

The first paper they published on this discovery, indeed the first paper that was ever published on what would soon be called fission—*Entstehung von Radiumisotopen aus Uran durch Bestrahlen mit schnellen und verlangsamten Neutronen* (The Production of Radium Isotopes from Uranium by Irradiation with Fast and Slow Neutrons)—appeared in the German journal *Naturwissenschaften* in late 1938.[12] As the title suggests, they were still reluctant to put in print what they already knew: that barium had been produced. As Hahn reports in his autobiography, while they were preparing this manuscript they tried to understand what had happened. In this, as Hahn himself admits, they made an almost incredible mistake in basic physics. At one point earlier in their relationship, Meitner once said to Hahn, *"Hänchen, geh' nach oben, von Physik verstehst Du nichts"* ("Hahn dear, go upstairs [to the chemistry laboratory], you understand nothing of physics").[13] My own view is that Hahn never understood the physics of fission. In any event, Hahn assumed here that the nucleus broke in two. Ignoring the initiating neutron the reaction is of the form, using the conventional notation with the atomic number at the lower right and the atomic mass at the upper left, $_{92}U^{238} \rightarrow _{56}Ba^{137} + X$, where X is the unknown partner in the fission. What Hahn assumed was that *mass* was conserved in this reaction so that $m_u = m_{ba} + m_x$ which would give $m_x = 101$, the element ruthenium. It is in the platinum group with atomic number 44. Two things are wrong with this. In the first place it is not mass that is conserved, but rather *energy*. The masses of the final products, when added up, are less than the mass of the parent, and the difference, by Einstein's equation, goes into the kinetic energy of the fission fragments. In the second place, the charges do not add up. You must have the sum of the charges of the fragments equal to the 92 of uranium. It is the *charge* that is conserved. In fact, if you compute $Q_u = Q_{ba} + Q_x$ you find $Q_x = 92 - 56 = 36$, which is the atomic number of krypton, a noble gas with an atomic weight of about 84 and thus chemically inert. Hence, the first nuclear fission observed was into barium and krypton.

As I mentioned, Hahn and Meitner were communicating during this time. She was as baffled as he was. If it had been anyone but Hahn, whom she knew to be a great chemist, she would simply have dismissed this observation as nonsense. She received Hahn's letter about it on December 21, 1938. She immediately wrote back: "Your radium results are very startling. A reaction with slow neutrons that leads to barium! . . . At the moment the assumption of such a thoroughgoing break up [of the uranium nucleus] seems very difficult to me, but in nuclear physics we have experienced so many surprises, that one cannot unconditionally say: it is impossible."[14] What happened in the next few days is one of the great stories of twentieth century physics. Meitner had been invited to spend the Christmas holidays with friends in Kungälv, on the west coast of Sweden. Her nephew Otto Frisch was also invited. Frisch's description of what happened is a classic. I will quote it in its entirety and then deconstruct it. He writes:

> When I came out of my hotel room after my first night in
> Kungälv, I found Lise Meitner studying a letter from Hahn, and obvi-

ously very puzzled by it. I wanted to discuss with her new experiments that I was planning, but she wouldn't listen: I had to read the letter. Its content was so startling that I was first inclined to be skeptical. Hahn and Strassmann had found that those three substances were not radium . . . [but] barium.

The suggestion that they might after all have made a mistake was waved aside by Lise Meitner; Hahn was too good a chemist for that, she assured me. . . . We walked up and down in the snow, I on skis and she on foot (she said and proved that she could get along just as fast that way); and gradually the idea took shape that this was no chipping or cracking of the nucleus but rather a process to be explained by Bohr's idea that the nucleus is like a liquid drop; such a drop might elongate and divide itself. . . . We knew that there were strong forces that would resist such a process, just as the surface tension of an ordinary liquid drop resists its division into two smaller ones. But nuclei differed from ordinary drops in one important way: they were electrically charged, and this was known to diminish the effect of the surface tension.

At this point we both sat down on a tree trunk, and started to calculate on scraps of paper. The charge of the uranium nucleus, we found, was indeed large enough to destroy the effect of surface tension almost completely; so the uranium nucleus might indeed be a very wobbly, unstable drop, ready to divide itself at the slightest provocation (such as the impact of a neutron).

But there was another problem. When the two drops separated they would be driven apart by their mutual electric repulsion and would acquire a very high energy, about 200 MeV [million electronvolts] in all; where could that energy come from? Fortunately Lise Meitner remembered how to compute the masses of nuclei from the so-called packing fraction formula, and in that way she worked out that the two nuclei formed by the division of a uranium nucleus would be lighter than the original uranium nucleus by about one-fifth the mass of a proton. Now, whenever mass disappears energy is created, according to Einstein's formula $E = mc^2$, and one-fifth of a proton mass was just equivalent to 200 MeV. So here was the source for that energy; it all fitted![15]

My explanation of this account is divided into two parts. This chapter explains the reference to the "liquid drop" and offers a qualitative explanation of the energy considerations. The next chapter contains more detail about the energy. The reason is that we need to have a more detailed understanding of exactly what is fissioning into what. The next chapter gives Bohr's argument that slow neutrons fission only the isotope uranium-235, a rare isotope that occurs at only about seven-tenths of a percent in natural uranium. It is the mass of this isotope that we need to compute the energy balance. Furthermore, uranium can fission in some 30 ways. Barium and krypton are not the most likely result by far. Hahn and Strassmann found this fission because they had a test that detected barium. During the war, they discovered many other fission modes. To see what energy will be liberated on average, we have to consider a more likely fission process. Moreover, the relatively heavy fission fragments are not the only result of fission. There are also neutrons, and we must discuss these as well. But let me, in any event, discuss the energy units. In chemical processes the "electron-volt" (eV) is the natural unit of energy. By the standards of daily use, such as illuminating a lightbulb, it represents a minuscule amount of energy. The effects are multiplied because myriad atoms are involved in chemical reactions. Typical nuclear reactions are in the few millions of electron-volts, and fission supplies 10 times more energy than that. A proton mass multiplied by c^2 corresponds to an energy of a billion electron-volts. That is the explanation of the last sentence in Frisch's account. Now to the liquid drop.

In 1928, the Russian-born theoretical physicist George Gamow (later the creator of the first Big Bang physics model), working in Copenhagen, proposed treating heavy nuclei such as uranium as manifestations of collective behavior. In a practical sense, it was not possible to describe the individual activities of some 200-odd neutrons and protons. This was in the spirit of how one would do the physics of a liquid drop. The myriad atoms would collectively produce an attraction—a surface tension—that would maintain the shape of the drop. Not much was made of this model until the 1930s, after the neutron had been discovered and people such as Heisenberg began seriously discussing the nuclear force. Bohr and his collaborators realized that a model like this was useful for treating nuclear reactions. A neutron, say, would be captured by a nucleus, which would result in what was known as a compound nucleus that in general would be left in an excited state. This nucleus would decompose itself by emitting a particle such as a neutron or an alpha particle. No one thought that it could decompose itself by splitting in two. For one thing, it was assumed that the surface tension would hold the drop together. Meitner was very familiar with the liquid-drop model. Weizsäcker, who had been Meitner's theoretical assistant for some time in 1936, used the model to derive what is known as the "semiempirical mass formula" for nuclei. What he did was to investigate the various effects that contribute to the nucleus's staying together, one of which is surface tension. Each one of these effects supplies an energy and hence a contribution to the mass. The amount by which this mass differs from simply the sum of the constituent neutron and proton masses tells us how tightly the particles are bound in the nucleus. This is the reference to the "packing fraction" in Frisch's description. Meitner must have known the Weizsäcker formula by heart. But she and Frisch realized that a real liquid drop and a nuclear liquid drop differed in a very significant way. In a nucleus there are two forces at work—an attractive nuclear force that binds together the neutrons and protons and an electrical force that acts repulsively among the protons. Like charges repel, and the more protons there are, the greater is this repulsive force. This is what makes the heavy nuclei unstable. The surface tension barely holds them together. When a neutron enters such a nucleus, the so-called drop fragments. In fact, for nuclei with a large enough atomic number, about 90, the nucleus fissions without being perturbed—spontaneous fission—a phenomenon that, as we see later, played a crucial role in the use of plutonium in nuclear weapons.

Frisch returned to Copenhagen. He told Bohr, who understood immediately and only wondered how no one had seen this before. Bohr was on his way to the United States with his collaborator Léon Rosenfeld. Bohr had agreed to say nothing until Meitner and Frisch could write up a paper. Frisch also wanted to check experimentally the existence of the energy pulses that would be produced by the fission fragments. This would be proof positive that fission had taken place, although without doing chemistry on what was causing the pulses one could not say what the fission fragments were. One possibility that he and Meitner had discussed was very heavy isotopes, which would be radioactive, of krypton and barium. This seemed plausible because of all the available neutrons from uranium that might be incorporated in such isotopes. In the meanwhile, Meitner and Hahn communicated by letter. This was a bit of a cat-and-mouse game. She did not want to tell Hahn what she and Frisch had done until it was published, but she wanted to give him hints that she was now sure that the barium was real. Hahn continued with his mistaken idea that mass, rather than charge, was conserved in the fission process. On the Bohr front there was a comedy of errors. He forgot to tell Rosenfeld not to say anything until Meitner and Frisch had published, so almost upon landing, Rosenfeld went to Princeton and told everybody. Practically by nightfall, fission had been confirmed in several laboratories. In those non–e-mail days, none of this was known to Meitner and Frisch, who continued their work at their own pace. It was not until the 16th of January that they submitted their two brief papers to the British journal *Nature*. The theoretical paper entitled "Disintegration of Uranium by Neutrons: A New Type of Nuclear Reaction," signed jointly by the two of them, appeared, along with Frisch's experimental paper, in the February 11th issue.[16] Frisch had asked a biology colleague in Copenhagen what term was used for cell division into two parts and was told that it was fission. This term was introduced in their joint paper in quotation marks, perhaps the last time quotation marks were used for it. It soon became the accepted term.

Copies of their papers had been sent to Hahn, and this was the first time he saw what they had been doing. Until a paper had been written and published, he could not refer to the work because it would show that he had been communicating with Meitner, which—in his now totally Nazified institution—he was afraid to reveal. Meitner and Frisch were careful to cite only published work so as not to compromise Hahn. It is not clear what his reaction was to this. He tried to maintain his priority for the discovery and to minimize what they had done to contribute to it. In the paper they wrote after they saw those of Meitner and Frisch, Hahn and Strassmann note laconically:

> During the writing of the reports on our latest experiments we received the manuscripts of two articles which are to appear in *Nature*, one by Lise Meitner and O. R. Frisch, the other by O. R. Frisch. We wish to thank the authors for sending them to us. Meitner and Frisch discuss in their manuscript the possibility of the fission of uranium and thorium [another element that Meitner and Frisch studied] nuclei into two fragments of about equal mass, for example barium and krypton. They discuss the possibility of such an event with regard to Niels Bohr's latest atomic model. O. R. Frisch reports on experimental proof of the formation of energetic fragments of the nuclei of uranium and thorium after neutron irradiation.[17]

There is nothing factually wrong with this statement, but what it doesn't say is that Meitner and Frisch were the first to understand what fission was. Hahn never conceded this, and it deeply hurt Meitner and probably was a contributing factor as to why she did not share the Nobel Prize.

In the next chapter, the scientific story resumes. Here, since Hahn and Meitner are together, I want to complete the arc of their lives and then end with a discussion of the aforementioned Nobel Prize. Hahn and Strassmann continued to work during the war. In the spring of 1945, Hahn was captured and sent to Farm Hall with the other nine detainees. Six months later the detainees were all returned to Germany, but forbidden to work on nuclear physics. Since Hahn was never a Nazi and since he had just won the Nobel Prize, he became something of an icon. He received all sorts of awards and honors. Perhaps the most interesting was the 1966 Enrico Fermi Award, which is sometimes given by the President of the United States. Hahn shared it with Meitner and Strassmann. He died in Göttingen in 1968, after a fall. As for Meitner, she was never happy in Sweden. She was treated as some sort of superannuated postdoc. She could never get the equipment needed to carry out the many fission experiments she wanted to do. Meitner told Hahn in a private communication that she resented his characterization of her as some sort of assistant and asked him how he would feel if she referred to him as her assistant. Hahn did not respond. Meitner remained in Sweden until 1960, when she moved to Cambridge to be close to her nephew. Frisch had spent much of the war at Los Alamos and then had taken a position at Cambridge. Meitner died in 1968, a few days before her ninetieth birthday and a few months after Hahn. Frisch died in 1979.

Ever since I learned about Hahn's 1944 Nobel Prize for chemistry (the actual award took place in 1945) "for his discovery of the fission of heavy nuclei," I have been curious as to how such a strange prize could have been awarded to Hahn alone a few months after Hiroshima. I knew that the Swedish Academy has a 50-year rule about giving out any records connected with the award of a Nobel Prize. I was also told that such information as there was was in Swedish and very laconic. Nonetheless, after the 50 years were up I tried to get these records but never succeeded. I was therefore surprised and pleased when in August 1996 an article appeared in *Nature* signed by Elizabeth Crawford, Ruth Lewin Sime, and Mark Walker, which was based on these records.[18] What they found was rather disturbing. The awarding of Nobel prizes in science is a three-stage affair. First there is a five-member committee representing the science in question, appointed by the Swedish Academy, which refers nominees to the relevant section of the academy, which then may refer the nominee to the entire academy.

In the case of fission, there was an interplay between the chemistry and physics committees. Select scientists external to the academy are asked to make nominations. In February 1939, the chemist Theodor Svedberg, who had won the Nobel Prize in 1926 and was chairman of the chemistry committee, nominated Hahn for a prize in chemistry. He knew of the paper of Meitner and Frisch but misunderstood its importance. By this time Niels Bohr and collaborators, especially John Wheeler, had produced a comprehensive theory of fission using the liquid-drop model. Svedberg believed that Meitner and Frisch had gotten their ideas from Bohr, rather than vice versa, so he dismissed Meitner's work. If Hahn had received the Nobel Prize that year, he would not have been allowed to accept it. Hitler had forbidden Richard Kuhn to accept the 1938 Nobel Prize for chemistry, and in 1939, Gehrard Domagk was forbidden to accept the prize for physiology and medicine and Adolf Butenandt for chemistry. Later, they were given the medals and diplomas but not the money. Until 1945, the chemistry committee had sole jurisdiction over fission, and they wanted to give Hahn the Nobel Prize. In 1945, the Swedish physicist Oskar Klein nominated Meitner and Frisch. Other members of the committee objected, renewing the claim that Meitner and Frisch had simply extended Bohr's work on fission. In September 1945, there was a move not to give a prize for fission at all because of Hiroshima, but this was voted down and Hahn was awarded the Nobel Prize in chemistry. It appears as if Meitner and Frisch were nominated after 1945, to no avail.

It is clear to almost everyone outside of Germany that this was an injustice. One of the German dissenters was Weizsäcker. He wrote a letter to *Nature*[19] justifying the awarding of the Nobel Prize to Hahn. Of course no one thinks that Hahn should not have gotten the prize. But so should Strassmann, who did a good deal of the work. They discovered a phenomenon that they did not understand until Meitner and Frisch explained it to them—and to us. In the next chapter, we discuss the real discovery of the transuranics, including, of course, plutonium.

VII
Transuranics

I cannot possibly describe either the excitement that this [the discovery of fission] produced in me or the chagrin I felt in realizing that I had failed to recognize this possibility myself on the basis of all the information available to me, information which I had studied so assiduously for a number of years. After the seminar was over I walked the streets of Berkeley for hours turning over and over in my mind the import of the news from Hahn's laboratory. I was in a combined state of exhilaration because of the beauty of Hahn's and Strassmann's discovery and disappointment because of my stupidity in not having recognized, myself, the fission interpretation of the wealth of experimental evidence available to scientists throughout the world.

Glenn Seaborg[1]

Bohr spent the spring of 1939 at the Institute for Advanced Study in Princeton, New Jersey. That year he and John Wheeler, who was at nearby Princeton University, produced their monumental paper on the liquid-drop model of fission.[2] I want to return to one section of this paper, an adumbration of a letter Bohr published in the *Physical Review*, which he had written the previous February,[3] that changed the technology of the exploitation of nuclear fission from that day to this, after I tell you a brief personal anecdote. In the late 1970s I taught a course in nuclear physics for graduate students for the first time. Needless to say, I covered fission. Being ambitious and seeing that my students seemed interested, I decided to go beyond the usual textbook treatment. In fact, we were using a textbook by Segrè, *Nuclei and Particles*,[4] which did the usual thing. There were a series of drawings, the first one of which was a sphere of positive charge representing the protons, which were supposed to be uniformly distributed in the sphere. Then the incident neutron, when absorbed, changes the shape of the sphere into an egglike shape that has the same charge.

The charge is determined by the number of protons, which throughout the process never changes. This transformation from sphere to egg starts the process off. It ends by the egg becoming totally elongated and breaking in two—fission. The first question was to find the conditions under which the initial step, the sphere into an egg, could take place. To this end, you had to calculate the electrical energy of the charged sphere, something one learns to do in a first course in electromagnetism. Next, you would calculate the energy of an almost spherical egg. For the process to take place, this energy must be less than that of the sphere. Systems normally evolve toward configurations of minimum energy. This was the problem. Segrè gave only the answer. If one assumed that this distortion would change the shape but not the total volume, the answer was incredibly simple. If d is a measure of the distortion, then for small d the energy diminishes by a factor of $d^2/5$. The $1/5$ was tantalizing. If the answer was so simple, why then did Segrè not give a derivation, which I was sure could be only a few lines? I then looked in all the textbooks I could get my hands on, and *nobody* gave the derivation, only the formula. This was really baffling.

As it happened, at that time I was interviewing Wheeler for what eventually became a profile for *Johns Hopkins Magazine*.[5] I decided to take advantage of this opportunity to ask Wheeler how one did the problem. Surely he would know. He said that he did not remember, but that it was in his paper with Bohr. I looked, and the answer was there without the derivation. Now things were serious. In those pre–e-mail days, I wrote and phoned several physicists whom I thought might know how to do this problem. I got various answers. One very distinguished British-born theoretical physicist said that it reminded him of a Tripos problem. The Tripos was a famous examination given at Cambridge to determine the "wranglers"—the brightest of the students. While this was an interesting piece of information, it did not advance the situation. I then wrote to Sir Rudolf Peierls, who was the head of the theoretical group at Oxford, where I had spent the previous year and had become convinced that Peierls knew everything. Indeed, he responded with a suggestion as to how to go about solving the problem. I enlisted a colleague at the Stevens Institute of Technology, where I was then teaching, the late Franklin Pollock, whom I knew to be a strong calculator. We rolled up our sleeves and actually produced two ways of doing the calculation, which we published.[6] Neither one was simple; certainly not something I would have given to my class.

I did have a chance to ask Wheeler how it was working with Bohr. Wheeler said that Bohr had "two speeds—not interested or completely interested."[7] As I can testify, when Bohr was not interested, his response was usually to say how interesting he found the talk—or whatever. If Bohr was actually interested, he would engage the speaker in a sort of gentle inquisition until he, and the speaker, understood whatever problem was being discussed. In connection with fission, the shoe was put on the other foot. There was at the Institute for Advanced Study that spring a brilliant, and extremely skeptical, Czech-born theoretical physicist named George Placzek. One day Placzek told Bohr that the liquid-drop model of fission was total nonsense. His reasoning was the following. The model requires an initiating neutron to start things off, to supply the energy needed to agitate the drop. On the other hand, it was a fact that as the rate of the fission reaction increased, the slower was the speed of the initiating neutron. Indeed, the reaction rate reached a kind of maximum when the neutron had no speed at all, and therefore no kinetic energy. Where then did the energy needed to agitate the drop come from? Bohr himself became very agitated and, with Wheeler in tow, began walking at high speed in random directions around Princeton. Then he had an epiphany, one that forever changed the practical uses of fission from reactors to bombs. Here is what he realized.

The nuclear force that holds neutrons and protons together in a nucleus is quite different from the forces with which we are familiar. For example, it must be a great deal stronger than the electric force because it keeps the nucleus from flying apart even though the protons are electrically repelling each other. Moreover, it is of extremely short range. Take the gravitational force for comparison. The Sun is 93 million miles from Earth, but its gravity is what keeps Earth in its elliptical orbit. By comparison, the nuclear force operates over such a small distance that, to keep from writing absurdly small numbers, a unit of length has been defined that applies to it. The unit is known as a fermi, and it is 10^{-13} centimeter—that is, 1/(1 followed by 13 zeros) = 1/10,000,000,000,000 centimeter. The range of the nuclear force is the order of a fermi. What this means is that in a nucleus the neutrons and protons interact only with their nearest neighbors. This property has important consequences in predicting which nuclei are likely to be stable. Nuclei are likely to be stable when neutrons and protons can pair off. The most stable of all are the lighter nuclei with the same number of neutrons and protons. Nuclei tend to cluster toward the line in which the neutron number N and the proton number Z are as close as possible. For light nuclei such as oxygen, which are extremely stable, the numbers of neutrons and protons are identical. In the case of oxygen there are eight of each. For the heavier nuclei, neutrons are added so that their contribution to the nuclear force can balance the repulsive electric force from the added protons. For a nucleus such as uranium-238, for example, there are 92 protons and 146 neutrons, which means that there is an excess of 54 neutrons. This has consequences for fission that I will return to once I have explained Bohr's epiphany.

To understand Bohr's epiphany, recall how fission works in the liquid-drop model. An incident neutron is absorbed by a nucleus, say uranium-238. It then forms a compound nucleus—in this case uranium-239. The question then is, How tightly bound is this nucleus? To understand why this is the question let me take a much simpler example. A single neutron can capture a single proton to form the nucleus of heavy hydrogen—the deuteron. The deuteron is less massive than the free neutron and proton. We know that mass is energy, so the excess mass is given off as energy. A gamma ray, a very energetic light quantum, is emitted. This light quantum carries the energy difference between the free neutron and proton and the deuteron. This energy difference is known as the binding energy of the deuteron. We can break up the deuteron into a free neutron and proton by irradiating it with a gamma ray of this energy, a process that is called the "photodisintegration" of the deuteron. Now we begin to see the outlines of Bohr's epiphany.

There is another thing we need to know. For light nuclei, the most tightly bound are those with the same numbers of neutrons and protons. As nuclei get heavier, there is an excess of neutrons. The most stable situation occurs when there is an even number of neutrons and protons, for example, in uranium-238, which has 92 protons and 146 neutrons, so no particle is without a partner. Nuclei with an even number of protons and an odd number of neutrons are less tightly bound. An example is uranium-235, which has 92 protons and 143 neutrons. The fact that it is less tightly bound indicates why there is so little of it in natural uranium—less than a percent.

Now let us look at fission. The compound nucleus for the fission of uranium-238 is uranium-239, while the compound nucleus for the fission of uranium-235 is uranium-236. But uranium-236 is, by the pairing off of the neutrons and protons, more tightly bound than uranium-239. Thus, when the neutron is captured by uranium-238, there is less energy given off than when it is captured by uranium-235. And here is the point. The liquid drop resists breaking up. You must give it enough of a shove for it to do so. Putting the matter less allegorically, you must supply a minimum amount of energy, what is called the "threshold energy." A neutron captured by uranium-235, turning it into uranium-236, gives off enough energy to get over this threshold, while a neutron captured by uranium-238, turning it into uranium-239, does not. Uranium-235 is therefore called "fissile," which means that neutrons of any energy can fission it. You can fission uranium-238, but only with neutrons that have sufficient energy. This was Bohr's epiphany.

Bohr outlined this result in his brief letter to the *Physical Review.* He was not thinking of applications, still less of nuclear weapons. But this enterprise began at about the same time. It was realized that to make a fission weapon out of uranium you would have to separate the isotopes. This process can be performed chemically for uranium only with great difficulty because the isotopes have, of course, the same chemical properties. You must use other physical methods, such as centrifuges, methods about which we still hear much. What about Placzek? He started all this, but he still thought it was nonsense. He and Wheeler made a bet. Wheeler bet $18.36 to a penny that it was right: The number 1,836 is the ratio of the proton-to-electron masses. When Bohr turned out to be right, Wheeler won. Placzek sent him a one-word telegram—"Congratulations!"—with a money order for a penny.

I said that I would discuss the implications of having a neutron-rich uranium nucleus involved in fission. To take an example, uranium-238 has 54 more neutrons than protons. In fission, where do these neutrons go? Most go into making neutron-rich fission fragments. These isotopes move to a balance between neutrons and protons by emitting electrons: beta decay. Most of the activity that people such as Fermi and his group and the Joliots, who thought they were seeing radioactive transuranics, observed came from the fission fragments. Additionally, however, neutrons are given off in the fission reaction. For uranium-235, on average 2.4 neutrons are emitted. This means that in many fissions two are given off, and in a lesser number three are given off. This is what makes chain reactions possible: These neutrons can initiate further fissions in which more neutrons are emitted in a cascade.

Figure 4 is the schematic of a typical fission in which two neutrons are given off; in this case, tellurium-137 and zirconium-97 are the heavy fission fragments. These are not the common isotopes but rather isotopes that have been made neutron rich. Now that we know which particles are emitted in a fission such as this, we can find the energy that is available to them. The simplest way to do this is to subtract the masses of the final particles from the masses of initial particles and multiply the result by the square of the speed of light. We are assuming that the neutron is captured at rest, so it has no kinetic energy. These days it is easy to find these masses. There are giant tables of the masses of all the so-called nuclides: the nuclei of elements with all their isotopes. Meitner and Frisch had to resort to somewhat cruder methods but got an answer—200 MeV—which is the right order of magnitude. Using such a table of nuclides I found that the reaction in Figure 4 yields 153 MeV, an enormous energy by nuclear reaction standards. Most of this energy goes into the kinetic energy of the fission fragments. Since these fragments are electrically charged, they make the large pulses, which is what Fermi missed but Frisch observed.[8]

The search for the transuranics can be divided into two eras: B.F. and A.F.—"before fission" and "after fission." The fault line between them was Fermi's (Plate 6) Nobel Prize in physics in 1938, "for his demonstration of the existence of new radioactive elements produced by neutron irradiation, and for the related discovery of nuclear reactions brought about by slow neutrons."[9] It seems as if Fermi was tipped off about the prize in advance so that he could bring his entire family to Stockholm in December of 1938, on their way to the

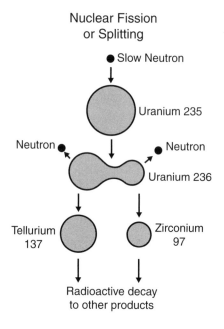

FIGURE 4 Nuclear fission or splitting.

United States, where he had lined up a job at Columbia University. His wife was Jewish and subject to Italy's racial laws. On December 12th he gave his Nobel lecture. I quote a paragraph:

Both elements [uranium and thorium] show a rather strong induced activity when bombarded with neutrons; and in both cases the decay curve of the induced activity shows that several active bodies with different mean lives are produced. We attempted since the spring of 1934, to isolate chemically the carriers of these activities, with the result that the carriers of some of the activities of uranium are neither isotopes of uranium itself, nor of the elements lighter than uranium down to the number 86 [radon]. We concluded that the carriers were one or more elements of atomic number higher than 92 [trans-uranics]; we, in Rome, used to call elements 93 and 94, Ausenium and

Hesperium, respectively [ekarhenium and ekaosmium]. It is known that O. Hahn and L. Meitner have investigated very carefully and extensively the decay products of irradiated uranium, and were able to trace among them elements up to the atomic number 96.[10]

But between the time the lecture was delivered and the time the written version appeared, everything had changed. Indeed, just after the quote above is a footnote clearly added to the original. It reads: "The discovery by Hahn and Strassmann of barium among the disintegration products of bombarded uranium, as a consequence of a process in which uranium splits into two approximately equal parts, makes it necessary to reexamine all the problems of transuranic elements, as many of them might be found to be products of the splitting of uranium."[11] The A.F. era had arrived.

The two significant things that made the real discovery of the transuranics possible were the discovery of fission and the invention of the cyclotron. I return to the latter momentarily. At the end of their brief paper, Meitner and Frisch spell out what to look for now that fission has been discovered. They write: "It might be mentioned that the body with the half-life of 24 min[utes] is probably really ^{239}U and goes over into eka-rhenium, which appears inactive but may decay slowly, with emission of alpha particles."[12] To spell this out, what Meitner and Frisch were saying is that uranium-238 can absorb a neutron and, instead of fissioning, become uranium-239. This isotope has a beta decay with a half-life of 23 minutes. Meitner, Hahn, and Strassmann had observed such a beta activity, but had not been able to identify where it was coming from. Such a beta decay turns a neutron into a proton and produces an element that is one step up in the periodic table. Meitner and Frisch in their paper are still calling this element "ekarhenium." It will turn out that its chemical properties have nothing to do with rhenium. They have also guessed wrong about the decay of this isotope—239—of element 93. It, in fact, beta decays into element 94—plutonium—of which much more shortly. Now to the cyclotron.

The cyclotron was the invention of Ernest Lawrence. He was born in 1901 in South Dakota and had earned his way through St. Olaf College selling pots and pans door to door. In 1925, he took a Ph.D. from Yale, and three years later was recruited by Berkeley to try to develop its physics department. A year later, Robert Oppenheimer was also persuaded to come to Berkeley. Between the two of them, they made the Berkeley physics department one of the best in the world. Lawrence was interested in trying to generate high-energy particles. Up to that time the standard method had been to accelerate charged particles in a straight line by subjecting them to electric fields. This method was limited by the distance over which the acceleration could take place, typically less than 10 feet. In contrast, the Stanford Linear Accelerator, founded in 1962, accelerates charged particles for some two miles. Lawrence had the brilliant idea of making the particles move in circles of ever-increasing radius. Figure 5 shows one of his early drawings. The two evacuated chambers on each side of the vertical gap are called, for obvious reasons, "dees." At the gap the charged particle is subject to an electric field that propels it into a dee. But the trick is to introduce a constant magnetic field at right angles to the motion. This field keeps the particle moving in a circle, because the particles are continually being deflected inward by the magnetic force. Lawrence made the very important observation from the theory

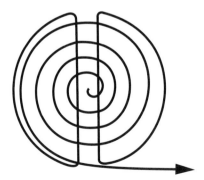

FIGURE 5 A sketch of a cyclotron showing the dees and a particle orbit in circles.

of such motions that the time it took for the particle to complete a circle was independent of the radius of the circle. Particles moving in circles with large radii had farther to go, but they were going faster. This meant that you could time the electric field so that it kicked the particle each time it came to the dee. Thus, you could get the particle to go faster and faster, the ultimate speed being limited by the size of the dees. Lawrence built the first successful machine in 1931 with a graduate student named M. Stanley Livingston. (This was his real name. I knew him.) The diameter of the two dees put together was about four and a half inches—a toy. The thing about Lawrence was that he was as efficient a promoter as he was a scientist. Even during the Depression he got money to build larger and larger cyclotrons. By 1938, the diameter of the dees had grown to some 60 inches. Enter Edwin McMillan.

McMillan, who was born in California in 1907, did his undergraduate work at the California Institute of Technology. In 1932, he got his Ph.D. in physics from Princeton and then went to Berkeley, where two years later he began working with Lawrence. By 1939, when the discovery of fission was announced, he was an assistant professor. He immediately began doing fission experiments, primarily to study the energy of the fission fragments. For this purpose he used the Berkeley 37-inch cyclotron. In it, he accelerated the nuclei of heavy hydrogen—deuterons, and these were made to collide with a beryllium target, producing an intense beam of neutrons that in turn were used to irradiate uranium. The study of the fission fragments did not lead to anything of great interest. However, uranium that had not fissioned, but rather was otherwise transformed, did lead to something new. There were two beta activities. One of them had a 23-minute half-life, which came from uranium-239, an activity that had already been discovered by Hahn, Meitner, and Strassmann. But there was a new beta activity with a half-life of 2.3 days. This, McMillan conjectured, came from element 93, the first transuranic. But now he needed to do the chemistry. Since he was not a chemist he asked Segrè, who was also not a chemist, to do the chemistry. Segrè agreed and got it wrong. In his 1951 Nobel Prize lecture (McMillan and Glenn Seaborg, whom we will come to shortly, shared the 1951 Nobel Prize in chemistry), McMillan said he had asked Segrè to collaborate because of Segrè's familiarity with the chemistry of rhenium. It was still held that element 93 was a chemical homologue of rhenium. What Segrè did correctly was to show that the 2.3-day activity did not come from anything with the chemical properties of rhenium. But what did it come from? Here is where Segrè got it wrong. Segrè came to the conclusion that the uranium had fissioned and that one of the fission fragments—the one with the 2.3-day activity—was a "rare earth." Rare earths (the name comes from eighteenth and nineteenth century chemistry) are neither especially rare nor are they "earths." They are metals whose abundance was somewhat less than other elements that were found in the earth. In a standard modern classification, rare earths are a set of elements in the row below the rest of the periodic table. It is the row that begins with lanthanum (57) and ends with lutetium (71). Segrè proposed that the unknown element producing the 2.3-day activity was actually a rare earth heavier than lanthanum. He concluded his letter to the *Physical Review*[13] on his finding (the italics are his) with the remark: "The necessary conclusion seems to be that the 23-minute uranium decays into a very long-lived 93, and that *transuranic elements have not yet been observed.*" His point being, and he states this explicitly elsewhere, that element 93 might even be stable, which is why the observed 2.3-day beta-decay activity must come from something else.

At first, McMillan accepted this result, but then he began to brood about it. It just did not seem to him that the 2.3-day activity behaved like a fission fragment. This was more of an intuition than anything else. At this point the 60-inch cyclotron was up and running, and it produced an even more intense neutron beam than the one he had been using. He was able to show that whatever was producing the 2.3-day activity barely moved after being created. Thus, it could not be a normal fission fragment, which always sped off at high speed. It must be something about as massive as the uranium itself.

He then decided that he had to do some chemistry. In this venture he had some good luck: the collaboration of Philip Abelson, who had fortuitously come to Berkeley on his spring vacation. Abelson had been born in Tacoma, Washington, in 1913. He had done his undergraduate work in chemistry at Washington State University and then had gotten a master's degree in physics, after which he had come to Berkeley for his Ph.D. He had then gone on to the Carnegie Institution in Washington, D.C., from which, in the spring of 1940, he was taking his vacation. It turned out that at Carnegie, Abelson had been trying to chemically separate the 2.3-day radioactive body from a large sample of uranium he had at his disposal. It became clear to the two researchers that they should collaborate.

By this time a great deal was known about the chemistry of uranium. In particular it was known that the uranium atom had several so-called oxidation states, configurations of atomic electrons that could readily participate in chemical reactions. (U(IV), for example, stands for uranium with four available electrons.) The two most important ones were U(IV) and U(VI), which had four and six available electrons, respectively. Fluorine is known as an oxidizing agent—indeed, the strongest—because of its ability to grab single electrons. This property results in two important uranium compounds: The first is uranium tetrafluoride, with four fluorine atoms grabbing single electrons, which is used in the manufacture of uranium hexafluoride (shown in Figure 6) with six fluorine atoms

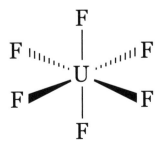

FIGURE 6 Uranium hexafluoride.

grabbing single electrons. Uranium hexafluoride is a highly cor-
rosive gas that is used in gaseous diffusion methods for separating
the isotopes of uranium. The notorious "yellow-cake" uranium
powder is transformed into this gas. One of the things that Abelson
and McMillan did was to see whether their unknown body reacted
with fluorine. It did, and they produced a tetrafluoride. By the time
Abelson's vacation was over, they knew they had a new element and
that it was a chemical homologue of uranium. They wrote up their
results in a brief letter, signed in the order McMillan and Abelson, in
the *Physical Review*.[14] As far as I know, McMillan and Abelson had no
further collaborations; Abelson returned to Washington. McMillan
did not follow this collaboration up either because, in November of
1940, he left Berkeley to go to Cambridge, Massachusetts, to work
on radar. This short letter won McMillan the Nobel Prize. Incidentally, in their note, element 93 does not have a name, but McMillan
had thought of one. Neptune is the next planet away from the Sun
after Uranus. So he called element 93 "neptunium." This act seems
not to have been revealed until August of 1945, when the so-called
Smyth report, *Atomic Energy for Military Purposes*,[15] was released.
In it, Smyth discusses the discovery of element 94, which is a decay
product of 93. For the first time he reveals that 93 was named
neptunium and 94 was named plutonium. I am going to turn to the
discovery of plutonium shortly, but first I want to discuss a remarkable physics paper by Maria Goeppert Mayer that was published in
the *Physical Review* in August of 1941.[16]

Maria Goeppert was born in 1906 in Kattowitz, Germany. Her
father became a professor of pediatrics in the university town of
Göttingen, which, because of its mathematics faculty, was considered
the mathematical capital of the world. There was in Göttingen only
one private school that prepared young women for the university,
and it closed down, because of German inflation, at the time Maria
was meant to attend it. But the teachers kept teaching anyway, and
Maria passed her examination for entry into the university. She had
intended to study mathematics, but instead became one of the bril-

liant theoretical physics students of Max Born, one of the founders of the quantum theory. Another of Born's students was Robert Oppenheimer. Goeppert led a small student rebellion to have Born silence Oppenheimer in classes, because he was doing most of the talking. She took her Ph.D. with Born in 1930, about the time she married the American theoretical physical chemist Joseph Mayer. She returned with him to the United States, where he had a professorship at Johns Hopkins University. She could never get a faculty job there but kept doing research in physics. In 1939, Mayer moved to Columbia University and Maria taught one year at Sarah Lawrence College. By then, Fermi had come to Columbia, and after McMillan and Abelson published their paper, Fermi suggested a problem that had to do with the fact that two neighboring elements in the periodic table, 93 and 94, seemed to have the same chemical properties.

To understand what the problem was, let us take a much simpler situation, the neighboring elements sodium and magnesium. Figure 7a depicts toy models of their electron structures. Note that in both cases we have a small number of electrons, one in the case of sodium and two in the case of magnesium, circulating around the same closed configuration of electrons. In fact, this is the configuration of electrons that is found in the neon atom. Indeed, it is customary to represent the configuration for sodium as $[Ne]3s^1$, meaning that there is one electron with no angular momentum in an energy state conventionally denoted by 3, circulating around a core of electrons identical to the electronic structure of neon. Recall that the ground state of hydrogen in the same notation would be $1s^1$. Likewise, the magnesium electronic structure would be represented by $[Ne]3s^2$—two outside electrons. While the depicted toy models are meant to be figurative, they do convey correctly the information that these outside, or "valence," electrons really circulate outside the core of the atom. This is why they get involved in chemical reactions. They can be shared with other atoms such as fluorine. Now we can begin to see the problem. With the transuranics we have two elements next to each other in the periodic table, 93 and 94, differing

sodium

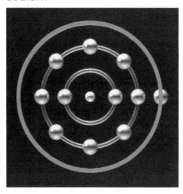

magnesium

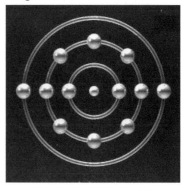

FIGURE 7a Sodium and magnesium atoms.

by one electron and yet having the *same* chemical properties. How is this possible?

In fact, this problem had already occurred. The lanthanides—the rare earths—show a similar characteristic. The elements across the lanthanide row all have sensibly the same chemical properties. The suggestion was made (really a qualitative suggestion without real theoretical justification) that what was happening was that the valence electrons remained the same but the shell levels below were filling up. To make this a little more visual, Figure 7b shows the toy models for neodymium (60) and promethium (61), two neighboring lanthanides. If you look very carefully, you will see that the two outside rings, the valence electrons, are the same in the two cases, but that in promethium an extra electron has been snuck into the next ring in. These interior electrons will not affect the chemistry. What Fermi wanted Mayer to do was to show that an arrangement like this was implied by quantum mechanics, which was, after all, the theoretical basis for the structure of the atom.

This is not the place to try to describe Mayer's paper in detail, but I would like to make some comments; first, on her plan of attack. A single atomic electron outside the closed core of electrons is subject

neodymium   promethium

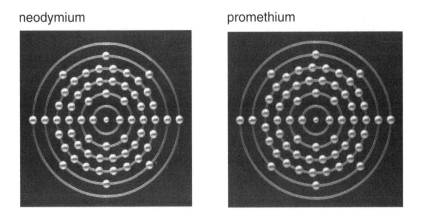

FIGURE 7b Neodymium and promethium atoms.

to two opposing forces:[17] an electrical force due to the protons that
tends to pull the electron into the nucleus and an opposing force
due to the fact that the electron in its orbit will, in general, have an
angular momentum. Thus, it is subject to a centrifugal force that
pulls it away from the nucleus. What happens is that there is a tug
of war between these forces. What Mayer did was to fix an angular
momentum and then move across either the lanthanide or the transuranic rows in the direction of increasing nuclear charge. What she
found was that at a critical value of the charge, the radius of the orbit
is dragged inside the atom. This is just what she was looking for,
since these electrons can no longer contribute to the chemistry. She
had found a quantum mechanical justification for the odd, related
chemistry of these elements. The next chapter, which discusses the
very strange physical and chemical properties of plutonium, contains
more about these electrons, but here let me make a couple of additional comments. By the time Mayer's paper was written, plutonium
had already been identified, but there is no mention of this in her
paper. It was a secret that was kept until the end of the war. It seems
likely that Fermi must have known about it, but he very likely did
not tell Mayer.

The second point I would like to make is about the impact of her paper. As far as I can tell, it had very little in fact, at least among people who might have found her result important. To take an example, Glenn Seaborg, the nuclear chemist whom we are about to meet, and who finally identified plutonium, got the idea in 1944, that the chemistry of the transuranics (he called them "actinides," since they apparently began with actinium, element 89) would be identical because an interior electron shell was being filled—exactly Mayer's point. He was then at the Metallurgical Laboratory in Chicago, part of the Manhattan Project, where knowledge of the chemical properties of plutonium was essential to its large-scale manufacture and use. He writes in his autobiography, *Adventures in the Atomic Age*,[18]

> When I broached the idea at upper-level Met Lab meetings, it went over like the proverbial lead balloon. I remember that at one meeting, the head of the chemistry division said that even if the concept was correct, he doubted it would be of much use. Latimer [Wendell Latimer, a senior chemist and a dean at Berkeley] told me that such an outlandish proposal would ruin my scientific reputation. Fortunately, that was no deterrent because at the time I had no scientific reputation to lose.[19]

Seaborg seems not to have realized that this concept was sanctioned by the quantum theory and accepted by no less an authority than Fermi. As for Mayer, after the war she and her husband followed Fermi to the University of Chicago, where they both became professors. Following another suggestion of Fermi, she began working on a problem in the structure of nuclei, for the solution of which, in 1963, she shared the Nobel Prize in physics. By that time, the Mayers had moved to La Jolla, California, where Maria died in 1972.

There is something prototypically American about Seaborg's life. He was born in 1912, in Ishpeming, Michigan, of Swedish immigrant parents. Indeed, Seaborg's mother tongue was Swedish. It was not an affectation when, in 1951, he gave the first part of his Nobel Prize address (he shared the prize in chemistry with McMillan) in

Swedish. His father was a machinist. In 1922, his mother decided she had had enough of small-town Michigan, and the family moved to California, just south of Los Angeles. Seaborg was educated in the public schools in Watts. He had no interest in science, but in high school he was required to take at least one science course to graduate. As it happened, the course being offered was chemistry, and with great good fortune he hit on an inspiring teacher, Dwight Logan Reid. Seaborg was 14. There and then he decided that he was going to be a chemist. But it was the Depression and his father had lost his job. There was no money for college, but here again he had the good fortune to live within commuting distance of the University of California at Los Angeles, UCLA, which charged no tuition for state residents. The summer before going there he worked as a chemist in a Firestone tire manufacturing facility, and as inexperienced as he was, he managed to save the plant from manufacturing a batch of defective tires. He showed sufficient brilliance at UCLA to be hired as a laboratory assistant, which gave him enough money to get through college. It was suggested that he go to Berkeley for his graduate work and he was given a job as a teaching assistant.

This was the time when Berkeley was building up its science departments with people such as Lawrence and Oppenheimer. The chemistry department also had its great man, the physical chemist G. N. Lewis, one of the most distinguished chemists of the twentieth century. Seaborg found himself more and more interested in nuclear chemistry, which meant using the cyclotron. He eventually did a thesis on the neutron bombardment of lead. Seaborg was now 25, and Lewis, who headed the department, asked him to stay on as his personal assistant. It was about this time that Seaborg first heard of fission and immediately began doing research on it. It was then that he first met McMillan, who had come to Berkeley from Princeton and would soon leave for the Massachusetts Institute of Technology (MIT) to work on radar. However, before he left, he had already started an experiment on uranium that seemed to show an alpha-particle decay that did not come from any of the known elements.

Seaborg joined the experiment, and in 1941, they produced, in McMillan's absence, a short note signed also by a graduate student named Arthur Wahl, which said that the alpha particles might be coming from element 94. The note was not published until after the war.

The work that had been done up to this time can be summarized in the following sequence of reactions. Here Np stands for neptunium and Pu for plutonium:

$$_{92}U^{238} + n \rightarrow \,_{92}U^{239} \xrightarrow[\text{23.5 min.}]{\beta^-} \,_{93}Np^{239} \rightarrow \xrightarrow[\text{2.33 days}]{\beta^-} \,_{94}Pu^{239}$$

In words, uranium-238 captures a neutron and then becomes the compound nucleus uranium-239. This beta-decays into neptunium-239, which in turn beta-decays into plutonium-239. The expectation was that this isotope of plutonium would decay by the emission of an alpha particle into uranium-235. This was the alpha-particle decay that Seaborg and Wahl tried to look for. However, it has a half-life of some 24,000 years, which meant its decay was so weak that it would hardly show up in a counter. They decided to try to produce a different isotope of plutonium that might decay more rapidly. To this end, they realized that they could use deuterons, the nuclei of heavy hydrogen—a neutron and a proton. Since the deuteron has a positive electric charge, it can be accelerated in a cyclotron, and thus a beam of high-energy deuterons can be produced. Deuterons can be made to impinge directly on uranium and the following reaction takes place:

$$_{92}U^{238} + \,_1H^2 \rightarrow \,_{93}Np^{238} + 2n,$$

where n stands for a neutron. This isotope of neptunium beta-decays into plutonium:

$$_{93}\text{Np}^{238} \rightarrow \frac{\beta^-}{2.1\ \text{days}} \rightarrow {}_{94}\text{Pu}^{238},$$

and this isotope of plutonium decays, with the emission of an alpha particle, into uranium-234 with a half-life of 80 years. With this enhanced activity, plutonium could be separated and the chemistry done. It was similar to the chemistry of neptunium and uranium. Seaborg, Wahl, and another collaborator, Joseph Kennedy, wrote a letter to the *Physical Review* with the understanding that it would not be published until after the war. Why the secrecy? It was evident to Seaborg that *mutatis mutandi*, the same argument that Bohr had used to show that uranium-235 was fissile, applied to plutonium-239. In fact, soon after its discovery, Seaborg and his colleagues performed an experiment on half a microgram of plutonium-239 using slow neutrons. They found that it was even more fissile than uranium. Plutonium was about to go to war.

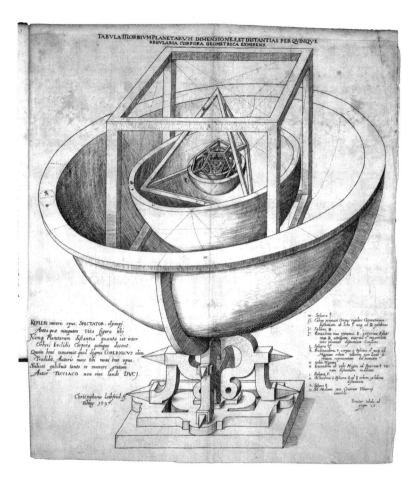

PLATE 1 The perfect solids.

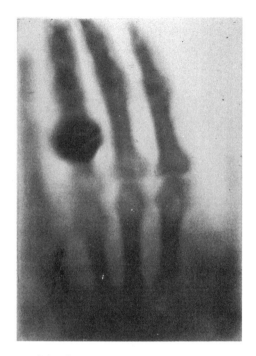

PLATE 2 Frau Röntgen's hand in x-rays.

PLATE 3 The Noddacks in their laboratory.

PLATE 4 Otto Hahn and Lise Meitner.

PLATE 5 Fritz Strassmann, 1902–1980.

Plate 6 J. Robert Oppenheimer, Enrico Fermi, and Ernest Orlando Lawrence.

PLATE 7 Manfred, Baron von Ardenne.

PLATE 8 Glenn T. Seaborg with a Geiger counter.

Plate 9 William "Willie" Zachariasen.

Plate 10 Cyril Stanley Smith.

Plate 11 Ted Magel, 1944.

Plate 12 Hanford site with the Columbia River, Washington.

Plate 13 Rocky Flats, Colorado.

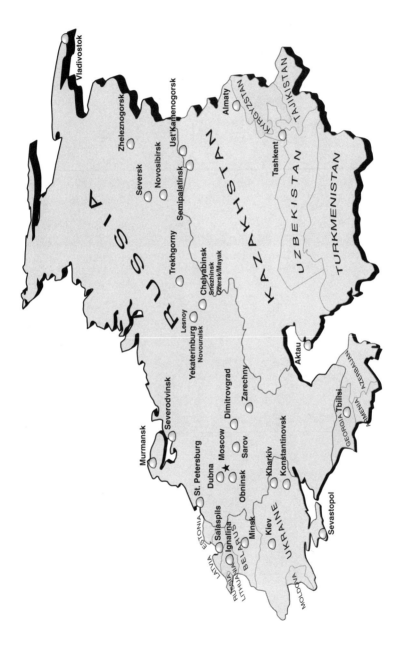

PLATE 14 A clear and present danger in Russia. The map shows the locations of stored plutonium.

VIII
Plutonium
Goes to War

As soon as such a machine [reactor] is in operation, the question of how to obtain explosive material, according to an idea of von Weizsäcker takes a new turn. In the transmutation of the uranium in the machine, a new substance comes into existence, element 94, which is very probably—just like $_{92}U^{235}$— an explosive of equally unimaginable force. This substance is much easier to obtain from uranium than $_{92}U^{235}$, however, since it can be separated from uranium by chemical means.

Werner Heisenberg, February 26, 1942, lecture to an audience that included high Nazi officials[1]

In the winter of 1938–1939, Louis Turner, a professor of physics at Princeton, took a sabbatical leave in Copenhagen. There he caught the fission virus from Bohr, who suggested an experiment that he and Frisch might do, but their equipment was inadequate to perform the experiment. In the spring of 1939, Turner was back in Princeton when Bohr and Wheeler were writing their paper. In January of the following year, Turner published the first review article about fission.[2] The list of references is interesting. Prior to 1939, there are some two dozen, but in 1939 alone there were many dozens. The transuranics get only a scant mention, merely to say that they have not yet been identified positively. However, after his review had been published, Turner was confronted by a puzzle. Why were the transuranics not as plentiful, say, as uranium on the surface of Earth? Presumably they had been created at about the same time and in similar quantities as uranium. What had become of them?

Turner proposed an ingenious solution. He suggested that at some time in the early history of the universe they had been exposed to neutrons and had fissioned. In particular, he proposed the fissionability of what he called $_{94}EkaOs^{239}$—ekaosmium. This was the first suggestion in the literature that plutonium might be fissionable. Turner was worried about publishing this, so he sought the advice of Leo Szilard. Szilard was a Hungarian-born physicist who in 1934, long before the discovery of fission, had the idea that some process involving neutrons might produce a chain reaction and took out a secret patent on the idea. After fission was discovered, Szilard realized its potential for making a nuclear weapon. In 1939, he wrote the letter that Einstein signed, addressed to President Roosevelt, warning of a possible German nuclear menace. Szilard had set himself up as a guardian of fission information, which is why Turner consulted him. Apparently, Turner's ideas were considered too speculative to be a security risk, so he published them.[3] However, Turner wrote a brief letter to the *Physical Review* entitled "Atomic Energy from U^{238}" in which he described how uranium-238 could produce ekaosmium.[4] He noted that the 239 isotope would be fissionable and thus a source of nuclear energy. Szilard objected to the publication of this note, and it was not published until 1946.

Incidentally, it turns out that one does not need Turner's rather baroque scenario to account for, say, the absence of terrestrial plutonium. All of the isotopes undergo alpha-particle decays with half-lives substantially shorter than the age of Earth. The longest-lived isotope of plutonium is plutonium-244 with a half-life of about

82 million years. There is no generally accepted evidence that any of it is still left.

In contemplating what happened at this time in the German and American programs, one can only wonder what might have happened if each side had known what the other was doing. I discuss the American program shortly, but first let me talk about the German. In September of 1939, German Army Ordnance (*Heereswaffenamt*) of the War Office created a program to develop nuclear energy, including explosives. People such as Heisenberg and von Weizsäcker were drafted to come to Berlin and work on it. They called themselves the *Uranverein*—the "uranium club." Heisenberg set about creating a theory of nuclear reactors, and some of the others began working on ways to separate the isotopes of uranium. On July 17, 1940, Weizsäcker delivered a five-page document to Army Ordnance concerning the use of ekarhenium as a fissionable element that might be useful as an explosive. At this time it had not been identified in the laboratory, so its physical properties were not known. But in June of 1940, McMillan and Abelson published their paper in the *Physical Review*, which came to Weizsäcker's attention after he had submitted his own report to Army Ordnance. I will explain how we know that in a moment. In their paper, element 93, which is what McMillan and Abelson called it, was shown to be unstable—beta-decaying into 94 with a half-life of 2.3 days. This made element 93, at least this isotope neptunium-239, useless for nuclear weapons, which Weizsäcker did not know at the time he submitted his July 1940 paper. (Neptunium-237 is a long-lived isotope that can be used in nuclear weapons.) But Abelson and McMillan had also argued that element 93 beta-decayed into element 94. They conjectured that they should be able to observe alpha particles from the decay of 94. When they didn't find them, they concluded that this decay must have a half-life of at least a million years. Actually, it is 24,000 years, but plenty long enough for the element to be useful.

It took some time for the *Physical Review* with the McMillan–Abelson article to arrive in Germany. After it did, Weizsäcker revised his proposal for using transuranics. This time he submitted it as a patent application to the *Reichspatentsamt*, the German patent office. There is no date on the copy of the patent I have, but from related material it must have been in the spring or summer of 1941.[5] The patent application is some six pages long. The first pages give a sort of history of fission. It is interesting because one can see what references Weizsäcker had available to him. The first reference is to Hahn and Strassmann, as one might imagine. Then there is a reference to Joliot and his collaborators.[6] They had demonstrated that neutrons were emitted along with the fission fragments, showing that a chain reaction was possible. That this was published in the open literature gave people like Szilard fits. Then Weizsäcker makes reference to a paper by a German physicist named Siegfried Flügge. Flügge published two papers, a technical one followed by a popular one, both written in the summer of 1939.[7] In the latter, Flügge pointed out that if all the uranium in a cubic meter of uranium oxide could be fissioned, the energy produced would be equivalent to what all of the German coal-powered generators produced in 11 years. This article caught the attention especially of the refugee physicists and helped persuade them that the German nuclear menace was serious. It also persuaded many of the German scientists to take nuclear energy seriously. Weizsäcker then refers to the paper of Bohr and Wheeler and to an extremely interesting one-page paper by Alfred Nier of the University of Minnesota with three collaborators from Columbia University, where the experiment I am about to describe was done.[8]

The goal of the experiment was to show that Bohr was right: The fissile isotope of uranium is uranium-235. The experiment is in two parts. First, one has to separate the isotopes uranium-235 and -238. This was done with what is known as a "mass spectrometer." The first step is to heat the uranium so that it becomes a vapor. Then this vapor is bombarded with electrons, which creates positively charged uranium ions by knocking some of the atomic electrons off. These ions are then subject to electric and magnetic fields arranged in such a way that only ions of a single fixed speed can pass a selector.

This beam of one-speed ions is now subject to a magnetic field that bends their motions into circles. One can show that the radius of the circle is proportional to the mass of the ion. If you have two isotopes, they will move in two circles, with the heaviest moving in the circle of the largest radius because it is harder to bend. Now the samples of the isotopes can be collected. The different isotopes, which have moved in different circles, arrive at different places on the collector. After roughly a half-day of running, the experimenters collected some tenths of a microgram of uranium-238 and less than a percent of this of uranium-235. I go through this detail to show how remarkable the experiment was. Then they took a beam of neutrons from the Columbia cyclotron, slowed them down with paraffin, and studied the fission of the uranium samples. They found that the fission of uranium-235 occurred at about the same rate as the fission of unseparated uranium. This showed that the fission of natural uranium by slow neutrons was all coming from the isotope uranium-235, as Bohr said it would. Wiezsäcker concluded from this result that if you wanted to build a *Uranmaschine*—a reactor—you would have to moderate the neutrons so that slow neutrons would be available to fission uranium-235. He suggested using heavy water—water atoms with deuterons and oxygen rather than protons and oxygen—as a moderator. He presumably understood that if you use ordinary water the protons will capture some of the neutrons, interfering with the fission chain reaction. He also remarks that to make a uranium nuclear explosive you would have to separate the uranium isotopes.

Nothing in any of this is patentable. Many members of the *Uranverein*, from Heisenberg on down, could have written the same summary. Heisenberg, for example, was designing a heavy-water–moderated reactor. However, it raises a question: Why were none of these experiments on uranium (Hahn and Strassmann excepted) done in Germany? The answer is that the Germans did not have a cyclotron, either then or at any time during the war, that they could use for any length of time. The nuclear physicist Walther Bothe of

Heidelberg tried to build one throughout the war. There is almost something operatic about his saga.[9] For example, the Siemens Company was supposed to make and deliver the magnet, which Bothe had ordered before the war. Siemens decided that the cyclotron did not have a high priority, so they held up delivery. Bothe sent his assistant Wolfgang Gentner to Paris to work with Joliot and learn more about the use of cyclotrons. Joliot had one, but he was in the Resistance and Gentner was an anti-Nazi, so nothing came out of that collaboration relevant to uranium. In1943, Bothe got his cyclotron running only to have to abandon it in 1944, when bombs began dropping nearby. After the war, the Germans were prohibited for a few years from working on nuclear physics. Bothe was a decent man, a German patriot and a non-Nazi. When he was captured he refused to give up any secrets until the war was officially over. Then he told what he knew. It was decided not to send him to Farm Hall. He and Max Born shared the 1954 Nobel Prize in physics, Bothe for his work in nuclear physics and Born for his work on quantum theory. Once the information from Allied physicists stopped coming to Germany, the German scientists were not able to learn any of the developments that followed the initial discovery of the transuranics.

After his summary of the history, Weizsäcker turned to the transuranics. It is here that he refers to the paper of McMillan and Abelson. This is the last bit of information on transuranics that he had access to for the rest of the war. He concluded that element 94, as he called it, could be manufactured in a *Uranmaschine* (reactor) and went on to describe its chemistry. He writes: "It's especially advantageous that the produced element 94 is easily separated (according to the rules governing ekarhenium or ekaosmium or similar rules) from uranium and can be produced chemically pure." This statement is entirely wrong, including the word "easily." But it is the basis of his patent proposal, which he motivates by describing the explosive power of element 94. There is no reference to the paper by Turner and, of course, no reference to Turner's unpublished letter. But Weizsäcker was a very good nuclear physicist who certainly understood that plutonium-239 would be fissile. The last two pages of Weizsäcker's document contain the six-part patent application. Basically what it says is that neutrons incident on uranium-238 can produce element 94, and to produce a significant quantity, a *Uranmaschine* is needed. The details of how such a machine is to be constructed are not given, He repeats his claim that by using known chemical processes (not specified), element 94 can be separated from the uranium matrix in which it was manufactured. Finally, he notes that element 94 can be used as a nuclear explosive. He does not indicate how such a bomb would be designed, or even how much of element 94—the critical mass—would be needed. The whole thing reads more like a research proposal than a patent application. One gathers that it was not acceptable to the patent office, because in June 1941, Karl Wirtz, who was in charge of the reactor project at the Kaiser Wilhelm Institute and was Weizsäcker's superior, upgraded it. There were then questions raised by the patent office about what exactly was being patented. On November 20, Wirtz responded, noting that what was significant was the suggestion of using element 94, which could be chemically separated from the uranium in a *Maschine*. Again there is no suggestion as to what the chemistry of 94 might be or how much would be needed. To call this an attempt to patent a plutonium bomb seems somewhat ludicrous. It would be as if Leonardo tried to patent the Boeing 747 on the basis of his drawings of flying machines. Weizsäcker and Wirtz had no idea of how to make an actual bomb and still less of an idea of what real plutonium was like. In the Farm Hall transcripts,[10] the British officer in charge reports that in the course of a discussion in August, Wirtz remarked that the Kaiser Wilhelm Institute had taken out a patent on an atomic bomb in 1941. Unfortunately, the people who were transcribing the recordings did not include what Wirtz actually said.

On August 9, 1945, the second nuclear weapon was dropped on Nagasaki. The Germans in Farm Hall discussed the event. The press and radio reports indicated that it was made of a different explosive material than that of the Hiroshima bomb. There was a conjecture, dismissed by Heisenberg, that it might be an element called "Pluto" which had been discovered, it was reported, in 1941. The reports even suggested that the element might have been made in a "machine." This, Heisenberg dismissed remarking, "I do not believe that the Americans could have done it. They would have had to have had, shall we say, a machine running not later than 1942, and they would have had to have had this machine running for at least a year and then they would have had to have had done all the chemistry."[11] I will come shortly to what the Americans did, but first I want to describe another German plutonium story that involves a character whose history is so bizarre that, if one put it in a novel, no one would find it plausible.

Friedrich ("Fritz") Georg Houtermans was born in 1903 in Danzig, now part of Poland.[12] His father was a wealthy Dutch banker, and his mother, Elsa, was related through her mother to the prominent Jewish Karplus family in Vienna. Houtermans went to live with his mother in Vienna. She had studied chemistry and biology and wrote a thesis entitled "Is Clean Water Dangerous?" Houtermans always had a spirit of contradiction, which he must have inherited. In later years it nearly cost him his life. He attended the *Akademische Gymnasium* in Vienna from which he was expelled after reading the Communist Manifesto in the lobby of the school on May Day, 1919. It was probably about this time that his mother sent him to Sigmund Freud to be analyzed. Freud also expelled him after Houtermans confessed that he had been making up his dreams. One might have thought that would have been a good subject for analysis. Somehow, Houtermans acquired enough discipline to study for, and pass, the entrance examination for the university in Göttingen in 1921, at about the same time that Meitner had gone there. He took courses with Born but worked mainly with James Franck, who won the Nobel Prize in physics in 1925. Franck ran the chemistry department in Göttingen until 1933. He was Jewish and, in 1933, emigrated to America, ending up during the war as director of the Chemistry Division of the Metallurgical Laboratory at the University of Chicago, Seaborg's destination. Houtermans got his doctorate in physical chemistry in 1927. One of the people whom he met in Göttingen was a young woman named Charlotte Riefenstahl, who had gone there in 1922, to study physics and chemistry. Oppenheimer had a "crush" on her. Given his then-adolescent relationships with women, this seems like the appropriate term. On some outing, Oppenheimer arrived carrying a very expensive leather suitcase. Both he and Houtermans had generous allowances from their families. She admired the suitcase, so Oppenheimer gave it to her. Nonetheless, in 1930, she married Houtermans in a ceremony witnessed by Pauli, among others. She kept the suitcase.

By this time, Houtermans had done some work in physics that had gotten him considerable recognition. In the 1920s the great English astrophysicist Arthur Eddington had declared that nuclear reactions could not take place in stars. They were, he thought, too cool even in their interiors. The atoms, stripped of their electrons as they would be in this environment, carry a positive charge from the nuclear protons. This sets up a repulsive energy between nuclei that, according to classical physics, as Eddington claimed, could not be breached by nuclei in stars. Because of the low temperature, they would not have enough kinetic energy to cross the electrical energy barrier. But then came quantum theory. In quantum theory you can violate energy conservation for extremely short times. This is one of the Heisenberg uncertainty principles. George Gamow realized that this property would explain the physics of alpha-particle decay. The alpha particle in the nucleus of such a decaying particle would also have to penetrate the electrical energy barrier set up by other protons. Classical physics implied that this was impossible because of energy conservation, but quantum physics allowed it. Gamow worked out the theory and it fit the data. In 1928, he was at Göttingen and gave a talk about this. In the audience were both Houtermans and a young English astrophysicist named Robert Atkinson. They realized that Gamow's "barrier penetration" could run in reverse so that two light nuclei could overcome the electrical barrier and fuse. If the result produced less massive particles, energy would be given off, and this energy could power the stars. The details of their paper are not correct, but the basic idea is. A decade later people like Weizsäcker formulated the correct theory. This sort of work gave Houtermans the reputation of being a brilliant young physicist, someone to watch. In the early 1930s, he became assistant to the physicist Gustav Hertz (who had shared the Nobel Prize with James Franck) in Berlin. They worked on isotope separation methods, not of course for any military reasons, but because they were interested in the science. Then came Hitler. Hertz, who had Jewish ancestry, was more or less hidden as an industrial physicist with the German firm Siemens. After the war, he finished his career in the Soviet Union.

Houtermans, apparently, was having too good a time to take the growing menace seriously. In addition to being Jewish, he was also an itinerant member of the Communist Party, not a desirable combination in Germany at that time. He always had very left-wing sympathies, which he made no attempt to hide. However, his wife took the situation seriously. At her insistence, he got a job in England at EMI ("His Master's Voice" was their record trademark). He liked neither his job nor England, and against the advice of people like Pauli who had been there, he took a professorial job in the Soviet Union in 1935 at the Ukrainian Physical and Technical Institute in Charkov. In 1933, when he and his wife were moving to England, Houtermans had nearly been arrested at the German border by the Gestapo because he was carrying a collection of left-wing newspapers. They recorded his name, something that came to play a role later.

When their luggage arrived in the Soviet Union from Britain in 1935, the Houtermans again had problems. The police wanted to know why it contained seven editions of the Bible (no one knows). By 1936, the menace of the Stalinist purges was becoming clear. For example, Houtermans was forbidden to mention the Heisenberg uncertainty principle in his lectures, and then people began to get arrested. In 1937, it was his turn. When his wife realized what had happened, she escaped, unknown to Houtermans, to Copenhagen and then to the United States. Remarkably, she became Mayer's successor at Sarah Lawrence. While doing research for this chapter, I discovered that I owed her a debt. When I was trying in the early 1950s to learn about the quantum theory of fields, there was basically one textbook, *The Quantum Theory of Fields*, by the Swiss physicist Gregor Wentzel. Charlotte Riefenstahl Houtermans was one of its English translators.

Houtermans spent the next two and a half years in prisons in the Soviet Union. How he survived, I cannot imagine. The second of the prisons to which he was transferred, the Butkyra, housed him in a cell that was built for 24 prisoners but now held 140. In January 1938, he was sent to a prison in Kharkov, where he was subject to a very brutal interrogation; in other words, he was tortured. After the twelfth day of this treatment, he was told that unless he confessed to being a German spy, his wife and children would be arrested and he would never see them again. He did not know that they were safely in the United States, so he agreed to confess, after which, for the next two years, he was simply left in prison. One of the things that helped him survive was that he did mathematics, number theory especially, in his head; indeed, at one point he thought he had found a proof of Fermat's last theorem.

From the United States his wife attempted to help him, even enlisting Eleanor Roosevelt, whom she had met. Whether this had any effect on his transfer to the Ljubjanka prison in Moscow, where he was treated better, is not clear. He was given writing material and was able to write out his work on number theory. In August of 1939, Hitler and Stalin made their pact. One effect that this had on Houtermans was that on April 30, 1940, he was brought to the border town of Brest-Litovsk in Belarus and turned over to the Germans. He was promptly arrested by the Gestapo. They still had his Communist affiliation on record. He was put in prison in Berlin. By this time, he had lost all his teeth. What happened next is almost beyond belief.

A fellow inmate in the *Alexanderplatz* prison was being released. Houtermans gave him a message to deliver to an old friend in

Berlin. The message read, *"Fissel ist in Berlin."* "Fissel" was one of Houtermans's nicknames. The friend, to whom the message was delivered, knew that this must mean that Houtermans was in jail somewhere in Berlin. He also knew who to contact for help—Max von Laue. Laue was a physicist who was born the same year as Einstein, 1879. Indeed, in 1905 when Einstein's paper on relativity came out, and von Laue was Planck's assistant, Planck encouraged him to go to Bern and see Einstein. He was surprised that he and Einstein were the same age. Laue was the first modern physicist that Einstein had met. Laue wrote the first text on relativity and, in 1914, won the Nobel Prize in physics for his work on crystals. He became a professor at the university in Berlin. Laue was the only distinguished scientist I know of who remained in Germany and was an outspoken, and defiant, critic of the Nazis. Why he did not end up in a concentration camp I do not know. When Houtermans left Berlin to go to England, Laue was the person who saw him off. A few years later he helped Meitner escape. Now he toured all the prisons until he found Houtermans, giving him money and food and arranging for his release. The question was what to do with him next. As usual, Houtermans did not take any pains to conceal himself. He casually published a brief paper in *Naturwissenchaften*, signing it and giving his new home address in Berlin. Laue must have realized that this was going to get Houtermans in trouble, so he found a job for him where he would be, more or less, out of sight.

In Berlin was an inventor and entrepreneur Manfred von Ardenne (Plate 7). He had invented some devices used in radio and television, and the patents had earned him a great deal of money from the German post office. With this, and with some additional support from the post office, he had created a private laboratory on his estate in a suburb of Berlin. Upon hearing of the electron microscope in 1938, he built an improved one. Laue came to the laboratory to use it. Paul Harteck, a physical chemist, a member of the *Uranverein* and the Farm Hall group, said in a 1993 interview, "I met him [Ardenne] once; he was a very bright boy. In his young years he started to make very good inventions. This impressed the Post Minister very much and he was given lots of money to develop his ideas." Commenting on the fact that, after hearing about fission in 1938, Ardenne decided to get into the field, Harteck added: "But I think that nuclear energy was a bit out of his line. He was to a certain extent a type much like Wernher von Braun. . . . They weren't what you would call scientists but used very modern technology to overcome all their difficulties with imagination and hard work."[13] Laue, incidentally, was one of the people who was sent to Farm Hall. This was as much for his own protection as anything else. Immediately after the war there were still fanatical Nazis known as "werewolves" who might have wanted to harm him. It was also thought that he might help reconstruct German postwar science. When he returned to Germany, he first went to Göttingen and then finished his career in Berlin. He died in 1960, when the car he was driving was hit by a motorcycle.

Exactly what Ardenne and his people did with respect to nuclear energy, and what their relation was to physicists such as Heisenberg and Weizsäcker and other members of the *Uranverein* is difficult to figure out. In Ardenne's autobiography[14] he gives his version of his laboratory's achievements. But Ardenne was a fabulist and one can never be sure what to believe. However, there are certain objective facts. At the end of this account of Ardenne and Houtermans's life story, I will describe what Houtermans did. It involves plutonium. As for Ardenne, he put much of his efforts into the separation of isotopes. He invented a version of what was known at the time in the United States as the Calutron. This was a classified electromagnetic separator of isotopes that Lawrence and his people invented. One might think of it as a dedicated cyclotron, that is, dedicated to isotope separation. It was employed at Oak Ridge, the facility in Tennessee, as a stage in the actual separation of uranium isotopes, which produced the material that was used in the Hiroshima bomb. How much uranium, if any, Ardenne's people actually separated I have not been able to determine. One thing is certain, however. When the Russians entered Berlin in the spring of 1945, they appeared to know where Ardenne's laboratory was. The Americans and British had already gotten hold of the *Uranverein.*

How the Russians knew about the German atomic scientists is an interesting story in its own right.[15] They had learned from espionage about the Allied project well before Hiroshima and apparently dispatched atomic search teams to Germany following their army, much as we did with the so-called ALSOS mission. This was an intelligence mission that followed the troops to learn about the German program. The Russians actually looked, unsuccessfully, for Hahn in the Kaiser Wilhelm Institute. Ardenne had conveniently placed notices in Russian that the institute was a scientific establishment. At the end of April, Ardenne received his first visit from the Russians. What happened next is again not clear. Whether Ardenne and many of his people willingly left for the Soviet Union, or were forcibly removed, is murky, although Ardenne did write a letter to Stalin offering his services. I do not know if he got a response. But they did go, along with their equipment, including the electromagnetic isotope separator. For the next nine years, Ardenne and his family lived a comfortable life in the Soviet Union. He was awarded a Stalin Prize of the first class and used the money to buy land for a private laboratory in East Germany. He also recovered the equipment that had been taken from Berlin. In 1954 he moved to East Germany. In 1997 he died in Dresden.

By the time the Russians arrived in Berlin, Houtermans was long gone. In 1944 he moved to the *Physikalische Technische Reichanstalt,* an institute devoted to research that was useful for industries. It had been evacuated from Berlin, because of the Allied bombing, to Ronneburg, a small town south of Berlin in Thüringia, which turned out later to have a uranium mine. In the summer of 1941, Houtermans had another of those adventures that make you wonder. It certainly made people who knew him wonder.

After the German army moved east into Russia, Houtermans joined some colleagues on a mission to Kharkov. The stated objective, or at least *his* stated objective, was to try to rescue one Konstantin Shteppa, a Russian historian, who had been a fellow inmate in the

Soviet Union. Needless to say, he did not find Shteppa, although after the war they wrote a book together under assumed names. But his choice of traveling companions was what caused comment. Among them was Kurt Diebner, who, as I have mentioned, was a physicist and a member of the Nazi Party. This, along with the fact that Houtermans was seen in Berlin wearing a military hat, persuaded some people that he had been a Nazi collaborator. I find this unlikely. I think he was just oblivious.

Houtermans was also oblivious to his absent wife, whom he divorced long-distance in 1943. He remarried the following year and fathered three children. In 1953, his first wife came back to Europe and he remarried her. This union lasted two years and then he divorced and married again. In 1945, he had taken a position in Göttingen, where he remained until 1952, when he became a professor in Bern. By this time he had again reinvented himself scientifically and was now working in geophysics to which he made important contributions. What always saved Houtermans was that people knew just what a brilliant physicist he was.

Houtermans was always a chain smoker. Finding tobacco hard to get in Ronneburg, he wrote a cigarette maker in Dresden on official stationery requesting a special Macedonian tobacco for an experiment on fog and smoke. This ploy worked once, but when he tried it a second time he was caught and kicked out of his job. That was when he went to Göttingen. In 1966, his cigarette smoking finally caught up with him and he died of lung cancer.

As entertaining as it is to write about Houtermans, he would not get into our story except for a report he wrote in 1941. Ardenne had given him the job of studying the theory of nuclear chain reactions. This report, which is one of four papers he wrote that year, is entitled *Zur Frage der Auslösung von Kern-Kettenreaktionen*—"On Questions of the Release of Nuclear Chain Reactions." It is dated August 1941.[16] There exist two nearly identical versions. The first one had a limited circulation. I am not sure if it was sent to anyone in the *Uranverein*, but the second version certainly was.[17] To understand why this occurred we need to explain briefly what happened to the *Uranverein*. In 1941, the army decided to stop supporting the program, because they did not see how it could help in the prosecution of the war. A civilian entity called the Reich Research Council took it over and Albert Speer became involved in financing it. Heisenberg was the de facto director, which enabled him to set the priorities of who would get scarce resources such as metallic uranium and heavy water. However, in 1944, Walther Gerlach was appointed as the head of all fission research done under the aegis of the Reich Research Council. This post apparently gave him the authority to redistribute Houtermans's report, which he did in October of 1944. The copy I have, with its warning *Geheim!* (Secret!), belonged to Harteck.

The text of the report runs some 35 pages; there is also a section of diagrams and a bibliography. The latter is interesting because of what it contains and what it does not contain. The papers stop in 1940. There is a reference to the paper of Bohr and Wheeler and to the review article of Louis Turner, but there is no reference to Turner's speculative papers on the fissionability of the transuranics or to the paper by McMillan and Abelson. Most of the report deals with the conditions for initiating chain reactions using either fast or slow neutrons incident on known elements such as uranium or thorium. Reading it gives one the sense that it was written by a first-rate physicist with a mastery of his material, but the last few pages are more speculative: Houtermans considers what could happen if uranium-238 absorbs a neutron. He is aware that the compound nucleus uranium-239 is unstable against beta decay and asserts that with a 23-minute half-life, it decays into what he calls "$EkRe^{239}_{93}$"— ekarhenium (i.e., neptunium). At this point he makes an interesting suggestion. It was already well known that the heavy elements decay in series, until they finally reached a stable element such as lead. Four series had been identified. They were characterized by a numerology that I can illustrate by the series that is relevant here. All of the atomic masses in this series can be represented by the formula $4n + 3$. Here, n is a positive integer. To see how this works, take uranium-235.

In this case we look for an n such that $4n + 3 = 235$, which gives $n = 58$. For uranium-239, the same calculation gives $n = 59$. What Houtermans proposed was that his ekarhenium was a member of this series and that it would beta-decay into an adjacent member of the series, which would be the mass number 239 isotope of element 94, that is, plutonium-239. He did not try to estimate the lifetime for this decay. (Unknown to him, Seaborg had observed the lifetime for this decay of 2.33 days.) Houtermans then asked what the properties of the product of this decay would be. He noted that, in general, there were two possibilities. It might be quite unstable, in which case it would decay in a series until it reached a long-lived element such as uranium-235. On the other hand, it might be stable, or nearly so, in which case you would have a fissile isotope of plutonium that was chemically distinguishable from uranium. I thought that he might use the theory of alpha-particle decay to estimate how long such an isotope might live, but he leaves it at that.

There is no mention of an explosive, but Houtermans was quite alarmed with what he had discovered. He felt he should warn the Allies that the Germans were now on the road to making a bomb. He tried this twice. The first was in a verbal message he sent in 1941, with a colleague Fritz Reiche, who was fortunate enough to be able to get out of Germany then. Whatever the intent, there is no trace of any effect this message had. In 1942, a cable was sent from Switzerland to the group in Chicago. It is said that it was sent at Houtermans's direction. How Houtermans knew about a group in Chicago, I do not know. The cable read, "Hurry up. We are on the track." This was presumably a reference to the fact that the Germans knew about plutonium. In actual fact, there was nothing to worry about. The Germans never had the remotest chance of making a plutonium weapon. To see what this really involved, I now begin to describe American efforts beginning with Seaborg and Chicago and, in the next chapter, Los Alamos.

I do not want to present anything like a detailed history of the atomic bomb project as it unfolded in the United States. But I do want to explain enough so that it will be clear why, in April of 1942, Seaborg moved to Chicago, rather than staying in Berkeley or going to Manhattan, to do weapons-related work.

Lawrence and his people were working on isotope separation in Berkeley, and Fermi and Szilard and their collaborators had been working at Columbia on the design of what became the first nuclear reactor. It went critical on December 2, 1942, in a squash court in the basement of Stagg Field at the University of Chicago. Einstein's 1939 letter to President Roosevelt, urging the development of the atomic bomb, was followed by another one in 1940. Neither had much of an effect on starting the program. The real impetus came in the spring of 1940, from the unlikely source of Frisch and Peierls, who were both at the university in Birmingham. The physics department there was headed by an Australian-born physicist named Mark Oliphant who had been a student of Rutherford. Oliphant had met Frisch in Copenhagen, and when it became clear that Frisch would have to leave Denmark because of the German invasion, he offered him a job. Peierls was already there in Birmingham, and Frisch moved in with Peierls and his wife, as did many celebrated physicists.

After the work with his aunt, Frisch continued experimenting with uranium fission. Among other problems on which he worked, he wanted to test Bohr's proposal that only uranium-235 was fissionable. To this end, he had done work on an isotope separation method that did not use a cyclotron and had been invented in Germany. In Birmingham, Frisch continued with his project. While he was waiting for a glass blower to produce an essential tube, Frisch began thinking about whether one could use uranium to make a bomb. The essential idea was clear to him. To make a bomb one would need a chain reaction that evolved very fast, in millionths of a second. He knew that the neutrons produced when uranium fissions moved at speeds approximately a tenth of the speed of light. This was promising, because the next fission in the chain could take place almost immediately. The problem was that if you used natural uranium, which is more than 99 percent uranium-238, these neutrons are not energetic enough to produce substantial fission, that is, enough to start a chain reaction. Thus, to make a bomb you would need uranium-235. But how much? It was at this point that he began his collaboration with Peierls.

The basic idea for computing this mass—the critical mass—is the following. Let us suppose, as Frisch and Peierls did, that somehow you have manufactured a solid sphere of uranium-235. How would you go about computing the critical mass of the sphere? First you would have to compute the critical radius; from that you could easily get the critical mass because you would know the volume of the sphere. If you multiplied this value by the density of uranium, you would have the critical mass. To find the critical radius you could reason as follows.

On the average, a neutron must travel a certain distance in uranium before colliding with and fissioning the next uranium nucleus. This distance is called the "mean free path" for fission. Clearly, if the radius of the sphere is smaller than the mean free path, neutrons will escape from the surface before they can cause fission. But if the radius is much larger than the mean free path, you are being wasteful of the nuclear material. Thus, the critical radius should be, in order of magnitude, about the same as the mean free path for fission. Now, the mean free path for fission depends on how probable fission is once the neutron strikes the next uranium nucleus. This probability depends on the cross section—the effective area of the target—for fission. The larger the cross section, the smaller is the mean free path. At this point, Frisch and Peierls were stuck. They did not know the cross section because they had no access to classified data, which in any case came from the United States. What they did know was that quantum mechanics fixes an upper limit for the size of the cross section, so they put it in. This, it turned out, was about a factor of 10 too large compared to the measured cross section. They put this maximum cross section into their calculation, which gave them a mean free path about a factor of 10 smaller than it should have been.[18] This underestimated the size of the uranium sphere. They got an estimated sphere about the size of a ping pong ball that weighed about 700 grams, less than two pounds. The present answer is that the critical mass for uranium-235 is about 123 pounds, while for plutonium it is considerably less.[19] I return to the plutonium case later because, like everything connected with plutonium, there are complications. The important thing is that this mass is a couple of pounds, not tons, which would have made nuclear weapons a practical impossibility.

Having reached this conclusion, the question was what they should do about it. Frisch realized that, although it would require a major industrial effort, separating this much uranium was within the realm of the possible. They felt they should inform someone. Indeed, they wrote two reports explaining what they had done. The problem was that neither of them was a British citizen. Indeed, they were technically classified as enemy aliens. They had produced something that if it had been done by someone else, they would not have had the right to see. They decided to confide in Oliphant. He was equally impressed and passed their report up the chain. In 1941, it became the basis of an official British document called the MAUD report,[20] which found its way to the United States and reinvigorated the program. By the fall of 1941, President Roosevelt was given a National Academy of Sciences report based on it about the possibilities for a fission weapon, and on January 19, 1942, he signed a brief note authorizing such a program. In between, there had been Pearl Harbor and we were at war. At this point the program was still under civilian control. It was not until September 1942 that the army, in the guise of General Leslie Groves, would take over and call all the shots. Until then, there was a very small council of scientist mandarins who ran the show. Among them was Arthur Compton, who had won the Nobel Prize in physics in 1927 and had for many years been a professor at the University of Chicago. He found himself overseeing groups on the East Coast, West Coast, and in between. In January of 1942, he decided that all the groups should be consolidated in Chicago. Thus, Fermi and Szilard moved from New York, and on

April 19, the day of his thirtieth birthday, Seaborg moved from Berkeley to Chicago.

By the time Seaborg made this move, there was a sense of urgency about the project. It was thought that the Germans had probably started theirs in 1939, with the discovery of fission, and might even be ahead. It was known that although many of the best scientists had been forced out of Germany, there were still some outstanding ones left. In particular, there was a fixation on Heisenberg, who it was conjectured must be deeply involved with the German program. When it comes to plutonium there is considerable irony about this. The Allies knew that the Germans were getting heavy water from Norway. The only use heavy water could have in a nuclear energy program was as a neutron velocity moderator for a reactor, which indeed is what the Germans were doing. But at no time during the war did anyone in the U.S. program put two and two together to conclude that the Germans must know about plutonium and had thought of making it in reactors.

Along these lines, a remarkable episode occurred at the end of 1943. It began the previous summer when a German physicist named Hans Jensen made a second visit to Copenhagen. On this visit he brought news of the German program. Jensen was not a member of the *Uranverein*—he was not entirely trusted because of his left-wing views. But he had contacts. He had got hold of something that he interpreted as the design of a German nuclear weapon. This is what he conveyed to Bohr. There was a drawing. No one whom I have contacted who saw it knows who made it. It might have been made by someone in the program, by Jensen, or by Bohr on the basis of what Jensen told him. In any event, the following September, Bohr escaped from Copenhagen, first to Sweden and then to England. Whether he showed the drawing to anyone in England I have not been able to learn. But then he came to the United States and there he certainly showed the drawing to General Groves.

Groves became alarmed and insisted that Oppenheimer stop everything and promptly convene a group at Los Alamos to examine the drawing. This meeting occurred on December 31 and Bohr was in attendance. So was Hans Bethe, who told me about it. Bethe said he took one look at the drawing and realized that it was the drawing of a reactor. He thought the Germans were crazy and planned to drop a reactor on London. It never occurred to him, or anyone else at the meeting, that the Germans knew about plutonium and were thinking of reactors as machines for making plutonium. As I have explained, it never occurred to the Germans, even after Nagasaki, that the Allies had the capacity to use reactors to make plutonium. One can only wonder what the effect would have been on either side if this had been understood. (Incidentally, Jensen shared the 1963 Nobel Prize in physics with Eugene Wigner and Maria Mayer.)

Glenn Seaborg (Plate 8) came to Chicago that April with a colleague named Isidore Perlman. They had been students together at UCLA and then Perlman had gotten a doctorate in physiology from Berkeley. He knew nothing about plutonium, but he seemed like a useful recruit. Indeed, at the time, he was the only recruit. Seaborg was put in charge of the nascent research, and they were assigned what had been a college student lab on the fourth floor of the Herbert A. Jones Laboratory. In his autobiography, Seaborg explains what their task was supposed to be:

> The chemistry group's challenge was to come up with a process by which we could separate out the plutonium from all the material in the aftermath of the chain reaction. The process would have to work on a large scale. The plutonium would be present in a concentration of about 250 parts per million. That meant that there would be about a half a pound of plutonium in each ton of irradiated uranium. The uranium would also contain a large selection of intensely radioactive fission products. So our challenge was to find a way to separate relatively small amounts of plutonium from tons of material so intensely radioactive that no one could come near; the separation [of the plutonium from the uranium] would have to be performed by remote control behind several feet of concrete. There could be no breakdowns

requiring repairs because the radioactivity would keep anyone from approaching the apparatus once it started operating.

We would have to develop this process for an element that now [in 1942] existed in such minute amounts that no one had ever seen it. All our knowledge of it was based on the secondary evidence of tracer chemistry—measurements of radioactivity and deduced reactions. Tracer chemistry was itself relatively new; deductions based on it were often subject to doubt.[21]

Seaborg's first task was to recruit other scientists; however, several handicaps stood in the way of this endeavor. In the first place, he did not have a significant scientific reputation. His greatest discovery up to that point was plutonium and this was a secret. In the second place, he could not tell a potential recruit what the project was until the individual was aboard and cleared. In his autobiography, Seaborg describes discussing the job in very vague terms with a chemist that he was trying to recruit. The chemist said that the details didn't matter since whatever it was it would involve the 92 known chemical elements. It was only later that he learned it involved element 94, which he had not known had been discovered. On top of this, Seaborg needed people who specialized in a discipline, extreme microchemistry with radioactive elements (which later came to be called "radioultramicrochemistry"), that at the time did not exist.

Despite these handicaps, Seaborg rapidly assembled a group. He must have been a very charismatic leader. The average age of the group was 25. No one was allowed to use the name plutonium, or even element 94, so they devised a code. The element was called "49," because it had an atomic number of 94 and an atomic mass of 239. This code, which was also applied to other elements (for example, uranium-235 became known as 25) was used throughout the war, even at Los Alamos. I leave it to the reader to decide the efficacy of this attempt at concealment.

Seaborg knew from the work he had already done at Berkeley that plutonium, like uranium, had stable oxidation states, that is, atomic states in which some number of electrons have been removed from the atom by putting it in contact with an element, such as fluorine, that grabs electrons. The relevant states for plutonium were the states in which four or six electrons had been removed. The question was how to exploit this property. In the first place one needed plutonium, which at the time came only from cyclotrons. The Berkeley 60-inch cyclotron supplied some plutonium, but most of it came from a cyclotron at Washington University in St. Louis that ran 24 hours a day for a year, irradiating uranium. Seaborg reports that in a year and a half the two cyclotrons produced two milligrams—the size of a grain of salt—of plutonium. Of course it came unseparated from the uranium matrix and the radioactive fission fragments. On one occasion, a truck pulled up from St. Louis with 300 pounds of irradiated uranium that had been packed in such a way that part of the sample had spilled out. The only advice that Seaborg could give to the people who were handling it was to wear rubber gloves and stay as far away as possible from the stuff. This sort of thing was fairly typical. They took enormous risks because they were convinced that they were in a race that could decide the outcome of the war.

There was no "cookbook" for the kind of chemistry needed to separate the plutonium—it was pretty much trial and error—but the basic idea was this: The uranium, plutonium, and fission fragment mixture that came from the cyclotron was dissolved in nitric and sulfuric acids. One of the chemists, Stan Thompson, whom Seaborg had rescued from a very tedious job at Standard Oil, had made an accidental discovery. He found that if he oxidized plutonium and put into the acid a high concentration of bismuth phosphate, the plutonium—in the oxidation state where the four electrons were missing—would attach itself to the bismuth phosphate and form an insoluble crystal that would precipitate from the solution leaving the rest of the detritus behind. Not only that, but it did so with a very high efficiency. The precipitate is then dissolved in nitric acid. An oxidizing agent is added, which bumps up the plutonium from a state in which four electrons are missing to a state in which six electrons are missing. Plutonium in this state does not precipitate with bismuth phosphate, and almost pure plutonium is left behind. It can be purified further by adding a reducing agent, which restores two of the electrons, and starting again. I go into this detail to show what was really involved, which contradicts what Wiezsäcker said naïvely in his patent application, namely that plutonium and uranium could "easily" be separated using conventional laws of chemistry.

By August 20 the Chicago team had isolated enough material so that, using a microscope, they could actually see a miniscule drop of plutonium. It was the first time a transuranic element had ever been seen by anyone. The next step was to delineate a process that could be scaled up a billionfold so that plutonium from a production reactor could be separated by the gram. Seaborg describes a meeting in August 1943 with Crawford Greenewalt of the DuPont Corporation. DuPont was going to build the production reactors, and Greenewalt had to decide then and there how the separation was going to take place. By this time, two methods had been discovered. One of them involved lanthanum fluoride; if it worked, it would produce a better yield and was chemically more understandable. On the other hand, if it didn't work there would be no yield at all. With bismuth phosphate, no one understood why it worked. In fact, James Franck argued that according to the conventional rules of chemistry it shouldn't work. But it did and produced a reliable, but lower, yield. With no hesitation, Greenewalt chose the bismuth phosphate, and on this basis the production reactors were built.

After the war Seaborg returned to Berkeley. Over the next few years he and his collaborators discovered new transuranics—for example, berkelium (97) and californium (98). In 1952, he was asked to be the faculty athletic representative and often traveled with various teams to games, and a few years later he became chancellor of the university. When President Kennedy was elected, he asked Seaborg to become chairman of the Atomic Energy Commission. In 1971, Seaborg returned to Berkeley where he took great pleasure in teaching freshman chemistry. He retired a decade later but was still active until close to his death in 1999.

In the summer of 1944, the first samples of plutonium with a mass of a gram or more started to arrive at Los Alamos from the production reactors. The problem was how to transform them from a laboratory curiosity into material for making a bomb. We turn next to that.

IX
Los Alamos

*Plutonium is so unusual as to approach the unbeliev-
able. Under some conditions it can be nearly as hard
and brittle as glass; under others, as soft as plastic or
lead. It will burn and crumble quickly to powder
when heated in air, or slowly disintegrate when kept
at room temperature. It undergoes no less than five
phase transitions between room temperature and its
melting point. Strangely enough in two of its phases,
plutonium actually contracts as it is being heated. It
has no less than four oxidation states. It is unique
among all of the chemical elements. And it is fiend-
ishly toxic, even in small amounts.*

Glenn Seaborg[1]

The few times I met him the late William "Willie"
Zachariasen (he died in 1979) reminded me of what I imagined a
Norse sea captain would be like (Plate 9). He had a look of bemuse-
ment, but gave one a clear feeling that he would bring the ship home
in high seas and in time for dinner. He was, in fact, the son of a
Norwegian sea captain and was born in 1906, in Langesund, Norway.
When he was young, he explored the islands in Langesundfjord

near his home. It was on these islands that Zachariasen first became fascinated by the crystallized rare-earth minerals that abounded there. While still in his teens he began his university studies with the great crystal chemist Viktor Goldschmidt. Goldschmidt was born in Switzerland in 1888, but had come to Oslo when he was 13. His father had become professor of chemistry at the University of Oslo. Willie used to row Goldschmidt out to a tiny island that Goldschmidt had bought for $500 in order to keep its crystals intact. Goldschmidt and Zachariasen collected those crystals.

Goldschmidt was one of the pioneers in the use of what is known as x-ray diffraction as it was applied to the study the structure of crystals. An x-ray is a very short-wavelength (about 10^{-8} centimeter) electromagnetic wave. Like all such waves, it can interfere constructively or destructively with similar waves. Von Laue, who won the Nobel Prize for this work in 1914, had been a pioneer in depicting such waves scattered from a lattice of atoms. They would show patterns of maximum and minimum intensities from which one could read off properties of the lattice. Likewise, a crystal is a periodic structure of atoms, the basic element of which is called the "unit cell." This cell is repeated again and again throughout the crystal. By studying the pattern of the intensities of scattered x-rays, an experienced observer can tell a great deal about the structure of the crystal, including the size of the unit cells. The term "cubic" means that the unit cell is shaped like a cube. Later, I exhibit how the unit cells look for plutonium, which is one of the patterns that Zachariasen discovered.

During the five years he spent on this work at the university, Zachariasen read hundreds of x-ray films with patterns like this one. He got his Ph.D. at age 22, the youngest person ever to have done so in Norway up to that time. Two years later, he received an invitation to take a position at the University of Chicago, where he spent the next 44 years. Before leaving Norway, he got married, and in 1931, his son Fredrick was born.

I first encountered Frederick Zachariasen in the late 1950s, when he was a postdoc at MIT and I was a postdoc at Harvard. It turned out that we were working on the same problem: the theory of deuteron photodisintegration. He went on to become a professor at Cal Tech, but we used to meet in the summers at the Aspen Center for Physics in Colorado. That is how I met his father, who came to Aspen for visits. From those meetings, I vaguely knew that he was a physicist and that he must be distinguished in his field, whatever that was, because he was both a professor and the chairman of the department at Chicago. This was a department that had people like Fermi in it. But I never asked him about his work. It is only when I was doing the research for this book that I came to learn that many of the insights we have about the structure of plutonium were originally due to him. Sadly, my colleague Fred died in 1999 of a heart attack.

As I have discussed in the previous chapter, beginning in 1942, Seaborg and his chemistry group at the Metallurgical Laboratory in Chicago (the "Met Lab") were working on methods of separating plutonium from the uranium in which it had been made, as well as separating the fission fragments and other impurities from the plutonium. They found, as noted, that bismuth phosphate would form crystals with plutonium and that these could be precipitated from an acid solution. No one knew why this worked. It was therefore decided that more had to be learned about the structure of bismuth phosphate, so Zachariasen was asked to join the project, which he did in 1943. Los Alamos was just getting started. It was proposed that it and Chicago divide up the work,[2] which meant an added assignment for Zachariasen. One of the tasks the Met Lab, and thus Zachariasen, was given was to use the micrograms of metallic plutonium that they had made with great difficulty to find out its density. Knowing this density was extremely important. The critical mass was highly dependent on the density of the material. For a solid sphere, for example, with the density ρ (the Greek letter "rho") and the critical mass M_c, M_c decreases as the square of the density; that is, $M_c \sim 1/\rho^2$, so that if you were to double the density by changing the material, or compressing it, you would need only a quarter as much to make a critical mass.[3] The problem was that there was almost no material to work with.

Before the production reactors began making plutonium, all of it was made in cyclotrons. In the fall of 1943, when it was decided to try to measure the plutonium density, Los Alamos had about 500 micrograms. The Met Lab asked to borrow it. Oppenheimer pointed out to James Franck, the Met Lab's director, in no uncertain terms, that the Met Lab already had two milligrams of the stuff, about four times as much, so it should make use of what it had. The person at the Met Lab who would make use of it was Zachariasen. Over the years he had mastered a technique known as "powder x-ray diffraction." As the name implies, this is the study of diffraction patterns from substances such as crystals that have been broken up into a powder. In such a powder the crystal unit cells are oriented randomly with respect to each other. This disorderly arrangement produces a very different kind of diffraction pattern, one in which the crystals of different orientations combine to produce a pattern of concentric circles.

An expert like Zachariasen could read such patterns. Indeed, it was a method that crystallographers had frequently used to identify the structure of crystals. He was allotted 100 micrograms of powdered metallic plutonium and set to work to try to determine its density. The idea was to find the size, and hence the volume, of a unit cell. The size was reflected in the diffraction pattern. The assumption was that these cells were densely packed together to make up the crystal. The atoms in the cells were predominantly plutonium, whose mass he knew, along with some trace amounts of impurities. Thus, he could find the mass of plutonium metal per cubic centimeter—the metallic density. The first value that he found was 13 grams per cubic centimeter, which can be compared to the density of water, one gram per cubic centimeter, or that of lead, $11^1/_3$ grams per cubic centimeter. So plutonium metal was very dense. But when he repeated the measurements, Zachariasen found values of 15 and then 15 $\frac{1}{2}$ grams per cubic centimeter. Zachariasen was not one to make mistakes like this. Something else had to be going on. What the Chicago group suspected seemed to confirm a fear that Seaborg had expressed sometime before, namely that there were impurities in the metal that might make plutonium useless as a weapon. To understand why, I have to explain a little more about bomb physics.[4]

Grosso modo, to make a fission bomb you begin with a subcritical assembly of material such as uranium-235 or plutonium-239 and, in one way or another, produce what is known as a supercritical assembly very rapidly, which then undergoes an explosive chain reaction. The devil is in the details of how you do this. I can illustrate with the kind of uranium-235 bomb that was dropped on Hiroshima, Little Boy. What we learn is relevant later when I discuss in some detail the kind of plutonium bomb that was tested on July 16, 1945, in Alamogordo, New Mexico, and then dropped on Nagasaki. Little Boy was what was known as a "gun-assembly" (Figure 8) bomb. In its unassembled state it consisted of two subcritical parts made largely of uranium-235, about 80 percent, and the rest of uranium-238. One part of the bomb was a projectile, and the other part was the target. The projectile was a stack of cylindrical rings about 10 centimeters wide and 16 centimeters long. This contained about 40 percent of the mass. The target, which contained the rest, was a hollow cylinder about 16 centimeters in both directions. The projectile could fit into the target. In Little Boy, the projectile was inserted into an antiaircraft barrel and then fired down the barrel into the target cylinder.

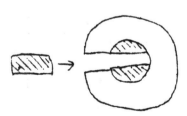

FIGURE 8 Serber's drawing of the gun assembly in his primer. This diagram is taken from the Los Alamos lectures given in 1943 by Oppenheimer's colleague Robert Serber. The lectures became known as "Serber's primer."

This is the only weapon of this design that was ever detonated.[5] It was never tested before it was dropped on Hiroshima. I introduce it here because we can use it to illustrate some important general principles. The first question is, What starts the chain reaction? Merely obtaining a critical mass is not enough. One needs an initiator—a source of neutrons. After a good deal of experimentation, the Los Alamos team came up with a design that involved polonium-210 and beryllium. Polonium is an alpha-particle emitter, and it was shielded from the beryllium until after the projectile hit the target. At this point, the alpha particles from polonium could impact the beryllium, which produced neutrons. The essential thing was that no substantial number of neutrons was produced before the activation of the initiator. Why this was essential is explained in what follows.

There are two regimes that are important for the explosion: the critical and the supercritical. They can be characterized in the following way. Each fission produces a certain number of neutrons. If you average these, it turns out that for uranium-235 the number is 2.52; for plutonium-239, it is 2.95. But all of these neutrons are not available for the next fission in the chain. Some, for example in the uranium bomb, get captured by the uranium-235 nuclei without causing fission, while others escape from the material through its surface. If we take the average number of neutrons produced per fission and subtract the neutrons lost to these effects, we end up with an effective number of neutrons that actually do cause fission. This number is usually designated by the letter k. If k is equal to one, the system is said to be "critical." There is always one neutron that is available to cause the next fission in the chain. Reactors are designed so that k is maintained close to one. On the other hand, if k is greater than one, the system is said to be "supercritical." Bombs are designed so that k is about two. I want to discuss this scenario, beginning with an explanation of how such a regime can be realized.

In both the plutonium bomb and the uranium bomb, you begin, as I have said, with the mass of the material in a subcritical state. The plutonium bomb is discussed more fully later, but the idea is that you compress the initial subcritical sphere of plutonium using high explosives. The explosion decreases the volume of the sphere, but does not change the total mass of plutonium. Hence, the density is increased. This means that the fission mean free path is reduced so that more of the neutrons fission than before, and fewer escape out of the material before they fission. Another way of saying this is that the critical mass is reduced. Since the actual mass remains the same, you have assembled, at this higher density, a supercritical mass. With the gun-design bomb, when the projectile hits the target the same compression effect occurs with the same results. Let me then suppose that using techniques such as this, we have managed to achieve a k of about two. What are the consequences?

The first thing I want to consider is, under these circumstances, how many generations in the chain reaction does it take to fission, say, one kilogram of uranium-235? By generations, I mean that with $k = 2$, the first fission generation produces two effective neutrons, each one of which fissions a uranium nucleus, producing four effective neutrons in the next generation, and so on. After n generations, 2^n nuclei have been fissioned. For uranium metal, which has a density of 19 grams per cubic centimeter, there are about 2.58×10^{24} nuclei in a kilogram. So we want to solve the equation $2^n = 2.58 \times 10^{24}$ to find the number of generations, n, needed to fission the entire kilogram. We can do this by taking the logarithm of both sides and solving for n. You will find that $n = 81$. Thus, with $k = 2$, it takes 81 generations to fission the entire kilogram. How long does this take? The speed of a neutron produced in uranium fission is about 1.4×10^9 centimeters a second. But the mean free fission path is about 13 centimeters. Thus, the time between generations is about $13/1.4 \times 10^{-9}$ second or approximately 10^{-8} second. The term of art for this unit of time, 10^{-8} second, in the bomb business is a "shake." So it takes about 81 shakes, less than a microsecond, to fission the kilogram. This timescale is important.

The passage from criticality to supercriticality itself takes time. In Little Boy, criticality was reached when the projectile and target

were separated by about 25 centimeters. To get a rough idea of the
time lapse involved until supercriticality, let us ask how long it took
for the projectile to move 10 centimeters. The speed of the projectile
in Little Boy was about 3×10^4 centimeters a second. Thus, to move
10 centimeters it took about a third of a millisecond. In Little Boy,
achieving supercriticality took a bit more than a millisecond because
the projectile had to move farther than 10 centimeters. The point
is that this is a long time compared to the microsecond it takes to
fission a kilogram when supercriticality is reached. It is essential that
no significant number of neutrons gets injected into the uranium or
the plutonium during this period. If this were to happen, there would
be a predetonation prior to the realization of supercriticality. There
would be what the bomb designers called a "fizzle." This would be
a very nasty explosion, but not the sort that flattened Hiroshima.
Where would such neutrons come from? In Little Boy, the uranium
isotopic purity—the relative amounts of uranium-235 as compared
to the residue of uranium-238 and uranium-234—was sufficiently
great (80 percent) so that the neutrons from the spontaneous fission
of these other uranium isotopes did not produce enough neutrons
to interfere with the effectiveness of the bomb. Hiroshima bears
testimony to that. But Seaborg's concern was, What would happen
with a plutonium bomb designed in the same way? I am not exactly
sure how he got his information about bomb design. This was prior
to the creation of Los Alamos. But Oppenheimer was already leading
a theoretical design study of these nuclear weapons and Seaborg was
certainly in contact with him. In any event, Seaborg realized that the
very way in which plutonium was created from uranium-238 would
generate impurities.

As we have seen, plutonium is generated in a two-step process
that begins with the capture of a neutron by a uranium-238 nucleus.
But there will also be some uranium fission going on. This means
that fission fragments will be created, but there are also impurities
in the reactor fuel, elements such as boron. Plutonium decays by the
emission of alpha particles. When these alpha particles collide with

an element like boron, neutrons are produced. The question Seaborg posed was, What concentration of these impurities could one tolerate so that the neutrons produced in such collisions did not generate a fizzle? His calculations showed that the concentration of impurity would have to be reduced to something like one part in 100 billion. Both General Groves and Oppenheimer were so informed. To make things worse, the British had realized the same thing, but their calculations showed that the impurity concentration would have to be reduced by another factor of 10. A discussion ensued of how such purity could be achieved, and the conclusion was reached only with great difficulty. Before anything was decided, samples of reactor-produced plutonium began to arrive at Los Alamos. As I explain shortly, these samples raised a new problem, which was of such a serious character that it definitely ruled out a gun-assembly plutonium weapon. The impurity question turned out to be, more or less, irrelevant.

Meanwhile, Zachariasen had begun the study of the structure of plutonium and its compounds that he would pursue throughout the war and, indeed, afterwards. In a report written in 1946, he noted, "For the past three years within the plutonium project, I carried out partial or complete crystal structure determinations of 140 different compounds of plutonium, neptunium, uranium, thorium or rare earth elements. My collaborator Dr. Rose Mooney made similar determinations of an additional 20 compounds."[6] Nearly all of the plutonium compounds were studied using x-ray diffraction, and even as late as 2005, Zachariasen's work on the structures of the oxides of these elements in these three years represented more than half of the total output of everyone else. As he later noted, "I remember working like hell on New Year's Day and all holidays; often I worked late for many, many hours to get the work done. I had a wonderful time. . . ."[7] One of Zachariasen's early discoveries was that plutonium has "allotropes." Allotropes are different crystal structures of the same element. The canonical example is carbon. Depending on how it has been treated, carbon can manifest itself, for example, as graphite or diamond. Allotropes are different from what we usually call "phases," which refer to whether the element is found in a liquid or a solid state, for example. Nonetheless, you will frequently find the term "phase" used for different allotropes. I also use it, from time to time, since I can't think of a better term.

The first two allotropes of plutonium that Zachariasen found were labeled with the Greek letters α and δ. Crystallographers label allotropes by Greek letters in order of the increasing temperatures at which the allotrope is question is stable. See Figure 11, which shows those allotropes and the temperatures at which they are stable. For plutonium, stability is a relative concept since it does not take much of a jar to cause an allotropic transformation. When he first discovered these allotropes, Zachariasen did not know there were four more. The full labeling is α, β, γ, δ, δ', and ε. We can worry about the rest later and concentrate here on α and δ. The first thing to emphasize is that an allotrope is not a property of a single atom. A plutonium atom is a plutonium atom is a plutonium atom. If you have seen one, you have seen them all. It is rather a property of the crystal structures that can be built out of these atoms. It is these structures that are, or are not, stable in a given temperature range.

Let's begin with the α-allotrope. It is stable up to a temperature of 122°C. This means that it is stable at room temperature. Zachariasen used x-ray diffraction to find the structure of its unit cell. While crystals can be exceedingly complex and unique—snowflakes, for example—they are nonetheless built out of a limited number of unit cell types. In the case of snowflakes they are hexagonal, which you can pick out in Figure 9a. The α-allotrope turned out to be "monoclinic"—a crystal structure in which all of the axes in the unit cell are not perpendicular to each other and may have different lengths. Figure 9a shows the 16-atom unit cell for α-plutonium. It looks perversely complicated. Moreover, it has less symmetry and hence little plasticity or pliability. Thus if you tried to bend α-plutonium metal, it would break like a piece of chalk. It behaves more like a mineral than a metal. On the other hand, the δ phase is quite

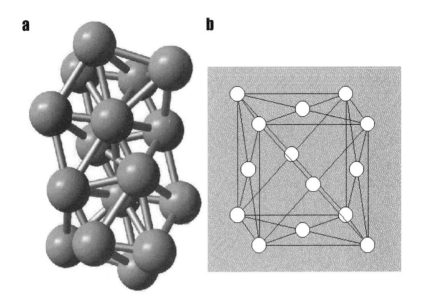

FIGURE 9a The 16-atom unit cell for α-plutonium.
FIGURE 9b The 14-atom unit cell for δ-plutonium.

something else. It is stable between 317° and 453°C and has a nice symmetric unit cell, what the crystallographers call a face-centered cubic (see Figure 9b). There are eight atoms in the corners and six in the center of each face, making 14 in all. We can imagine displacing this structure along a plane and preserving it. Indeed, δ-plutonium is as malleable as an ordinary metal, perfect for making into a bomb, except for the fact that at lower temperatures it readily morphs into the α phase, presenting a much greater engineering challenge. The densities are interesting. At 25°C, the α-phase density is 19.86 grams per cubic centimeter—very dense indeed—while at 320°C, the δ-phase density is 15.92 grams per cubic centimeter. The strange results that Zachariasen first found for the densities are explained by the mixture of different phases. Clearly, if you intend to use metallic plutonium to make a bomb, you will be confronted with a very significant metallurgical challenge. But worse is to come.

DuPont had been contracted to construct the power reactors. After Seaborg pointed out the impurity problem, there was some reluctance to proceed. However, once General Groves had decided to do something, it was next to impossible to stand in the way. Thus, beginning in February 1943, construction started on a pilot project located near Clinton, Tennessee—what later became Oak Ridge. It was designed to use the bismuth phosphate method of separation that had been developed at the Met Lab. The Oak Ridge reactor went critical in November and by April 1944, it was shipping grams of plutonium to Los Alamos; but it soon became clear that a disaster had occurred. To understand the issue let us review how plutonium is produced in a reactor.

The basic fuel in these reactors was natural uranium, more than 99 percent uranium-238, the rest being mainly the fissile isotope uranium-235. To enhance fission reactions, the neutrons created in fission are slowed down by a moderator—in this case, highly purified graphite, the same moderator that Fermi had used in his reactor. But some of the neutrons are absorbed by uranium-238 nuclei, producing neptunium-239, which beta-decays to plutonium-239. To get a substantial yield of plutonium-239, the reactor must be allowed to run for a reasonable amount of time. The longer the reactor is allowed to run before plutonium is separated from uranium, the more plutonium you get. However, while plutonium-239 remains in the reactor, it can absorb another neutron and become plutonium-240; but this isotope of plutonium spontaneously fissions, producing fast neutrons. There is now a balancing question, How much plutonium-240 can you tolerate without producing a weapon that will predetonate?

The fact that plutonium-240 would be produced was already known from the cyclotron production of plutonium. However, there was so little material to work with that measurements of the occurrence of this isotope were ambiguous. But now there were gram quantities, and Segrè was given the job of measuring the rate of spontaneous fission caused by the plutonium-240 in the sample they had. By late spring, Segrè reported that the spontaneous fission rate for this sample was at least five times as high as had been observed for the cyclotron-produced plutonium. By July 4 it had become clear that the gun-assembly method was not going to work for plutonium. It was just too slow. Neutrons would trigger a chain reaction before the material became supercritical. There was also a spontaneous fission issue for uranium-238, but in a bomb like Little Boy, some 90 percent of the material would be uranium-235, which had a spontaneous fission rate that was some 35 times lower. This is why the gun-assembly method worked for uranium. There was no realistic way of separating plutonium-239 from plutonium-240. They differed by one mass unit, while uranium-235 and uranium-238 differed by three, which makes a huge difference when you are trying to separate isotopes. My guess is that if the people working on the bomb had not been persuaded that they were in a desperate race with the Germans, and if General Groves had not shared this obsession, the project might have stopped right there and then. As it was, Oppenheimer got discouraged and considered resigning as director of Los Alamos. He didn't, but now the laboratory faced up to the two problems: metallurgy and assembly. I will begin with metallurgy. Enter into our story Cyril Stanley Smith (Plate 10).

Smith was born in Birmingham, England, in 1903. He got a degree in metallurgy from the University of Birmingham in 1924 and then a doctor of science from MIT in 1926. A year later, he began working at the American Brass Company in Connecticut's Naugatuck Valley. There he remained until the war, at which time he went to work for the War Metallurgy Committee in Washington, D.C. In February of 1943, while attending a meeting of the American Institute of Mining, Metallurgical, and Petroleum Engineers in New York, he was approached by the chemist Joseph Kennedy, who was one of Seaborg's collaborators in the discovery of plutonium and had been recruited to go to Los Alamos to head up its newly formed chemistry department. It is not clear why Kennedy contacted Smith in particular, although Smith had published a substantial amount of work and held several patents. It is also not clear what Kennedy could have told Smith about what would be going on at Los Alamos because Smith had no clearance. However, he told him enough, so that Smith saw going to Los Alamos as a way of escaping a desk job in Washington for which, as he later recalled, he had a "general distaste."[8] Not long after Smith's encounter with Kennedy, Oppenheimer had a recruiting talk with Smith on a park bench in Washington. Oppenheimer was very good at this sort of thing. By March of 1943, Smith was among the first group of scientists at Los Alamos. He was put in charge of creating a metallurgy group, without a clear idea of how big a job this was going to be. His first job was to find metallurgists who were not otherwise engaged in the war effort. This was not an easy task, but by 1945 when the war ended, he was running a department with 115 people in it. One of the difficulties in recruitment was that neither Smith nor anyone else, in the beginning, knew what such a department was supposed to do.

It was decided that the Los Alamos metallurgical group would not work on plutonium until gram samples arrived from the reactors, so they did various odd jobs, such as studying the properties of compounds of uranium with hydrogen. One of the oddest arose out of a request to take 620 pounds of gold and cast it into two hemispheres. Later Smith found one of the hemispheres being used as a doorstop. Once the plutonium began arriving at Los Alamos in half-gram lots in March 1944, the work to make a usable metal of it began in earnest. The first assumption was that its chemistry must be like that of uranium, because by this time it was understood how to make uranium into a metal: You began with uranium tetrafluoride (UF_4) and took advantage of the fact that if you heated it in the presence of calcium (Ca), the calcium would be more attractive to the fluorine than uranium would and you would induce the reaction $UF_4 + 2Ca \rightarrow U + 2CaF_2$, leaving uranium metal and calcium difluoride. Calcium here has acted as what is called a reducing agent. This sort of reaction was the way in which the micrograms of metallic plutonium that Zachariasen had been using had been made. This work was being done at the Met Lab by two young metallurgists Ted Magel (Plate 11) and Nick Dallas. At the end of 1943, Magel and Dallas were producing one-gram buttons of pure uranium metal from uranium fluoride. By early 1944, Oppenheimer had persuaded the Met Lab to relinquish Magel and Dallas, who arrived in Smith's group bearing their Met Lab equipment, which included a centrifuge. They had performed the uranium reduction in a centrifuge, which would then separate out the metal. They planned to do the same thing for plutonium. Later it turned out that a better idea was to use what was known as a stationary "bomb," a crucible specially lined so that it could contain the plutonium compounds. But Magel liked the thrill of the centrifuge; Smith referred to this approach as "excited, energetic, but slightly slap-dash."[9] Prior to the arrival of Magel and Dallas, the Los Alamos people, using their experience with uranium, tried to reduce plutonium trifluoride with calcium. They got what has been described as a "grayish cokey mass containing no agglomerated plutonium."[10] Then Magel and Dallas got into the act.

Magel's 1995 description of what occurred may be in the *se non è vero è ben trovato* category, but it is very amusing to read.[11]

> The reduction of a gram quantity of plutonium was considered a very big deal, because that amount of metal would allow much improved measurements of many crucial material properties. The reduction was supposed to take place on March 24, 1944, and General Groves and several top administrators had been specially invited to observe us as we did it.
>
> Well, when does everything go wrong—when you have a whole lot of observers, right? So on the 23rd I said to Nick [Dallas], "Let's go up to the lab and make the reduction tonight before all these people get here." Nick agreed, and we carried out the reduction using the hot-centered centrifuge bomb method. When it was done, we cut open the bomb, dropped the little button of plutonium metal in a glass vial and put it on Cyril Smith's desk with a note that read:
>
> *Here is your button of plutonium. We have gone to Santa Fe for the day.*

Everyone was pretty mad at us and claimed that we had contaminated the lathe and the back shop, when we opened the bomb
to retrieve the plutonium button. I don't believe that we had, but I
understood how they felt. In any case, once they had the button, they
immediately started measurements of the density and so forth. . . .

Magel and Dallas had produced the first sample (Figure 10) of
metallic plutonium that could be seen without the aid of a microscope. It enabled measurements of the allotropic phases. Figure 11
is a modern diagram, which is standard in any recent treatise on
plutonium. It shows the six allotropic phases previously referred to
as a function of the temperature at which they are stable. The temperatures are given in the Kelvin or "absolute" scale. To convert to
centigrade you just subtract 273.15. On the left-hand axis, atomic
volume is plotted. In the next chapter, when I discuss the science of
plutonium, I explain what atomic volume means. It is not as obvious

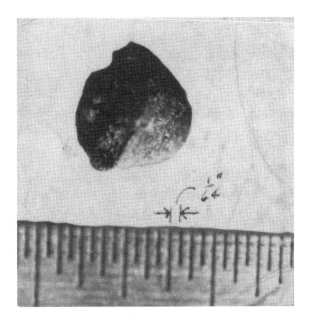

FIGURE 10 The first gram.

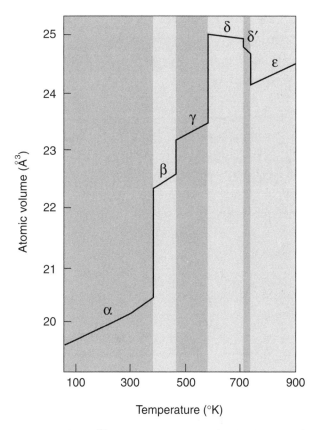

FIGURE 11 Atomic volume ($Å^3$ versus temperature (K) for the allotropes of plutonium labeled by the Greek letters. For nuclear weapons the important allotropes are the alpha and delta.

as one might think. The thing to note here is that there are two phases, $δ$ and $δ'$, in which the volume *decreases* when the temperature is raised. This is totally counterintuitive and is another example of just how bizarre an element plutonium is.

There were diagrams like this during the war but they were rather rough-hewn. Most of the details were filled in after the war. To give some idea, in 1958 a Russian chemist named Eugenii Makarov published what became a standard text on the crystal chemistry of uranium, thorium, plutonium, and neptunium. It was translated and published in the United States the following year.[12] Even in 1958, as their text makes clear, the crystal structure of the β phase was still unknown. The next year, Zachariasen showed that it had a similar structure as the α phase. During the war, anything that involved simple scientific curiosity was put aside if it did not contribute to making the bomb. One of the essential things that using these gram samples of metallic plutonium enabled the Los Alamos people to do was to measure the melting point of plutonium: the temperature at which the metal melts. The early Met Lab experiments gave results that seemed to be consistent with the kind of temperatures one finds for other metals. To give a few examples; iron melts at 1510°C, while steel melts at 1370°C, copper at 1083°C. On the other hand, it was discovered that plutonium metal melts at 640°C, an extraordinarily low temperature, which you had better know if you are going to use plutonium metal for something.

After making the first gram, Magel and Dallas made eight more grams of the superpure plutonium that was thought to be required for a gun-assembly weapon. But once it became clear that such a weapon was impossible, superpure plutonium was no longer needed, so they were out of a job. They decided to leave Los Alamos and join a small Manhattan Project group at MIT. In a 1995 interview, Magel was asked whether one of the reasons he left had to do with health. Magel was among the original members of the UPPU club—You Pee Plutonium. These were people who had had enough exposure to plutonium that it showed up in their urine. Magel recalled:

> Within weeks of making the first 1-gram button, I had an incident in which I was working in a dry box [a partially enclosed box, into which the hands can be inserted, designed to minimize plutonium's contact with the atmosphere] scrapping slag from another of those 1-gram buttons, and the needle I was using slipped, went through the rubber glove, and embedded in my finger. . . . I could see some black stuff in my finger. OK, I thought that's plutonium. . . . We went to the hospital and they thought they had cut it all out, but they hadn't—I still have some plutonium in one finger. They began taking

urine samples in 1945, which was when the procedure for measuring excreted plutonium was first available.

Magel continues:

Sometime between March and July of 1944, they developed a method of monitoring how much plutonium we were getting from breathing. The nose counts were the primary method for that. This girl would come around and swab our nostrils. One time I was getting ready to do a reduction, and I decided to take a last quick look inside this little tiny crucible to make sure I had put all the ingredients into it. I bent down close to it without bothering to put on my respirator. Apparently I got a very high nose count from doing that. But the big dose was from the needle stick. Dr. Voelz told me recently that I have the fifth-highest dose of the 26 members in the UPPU club.

George Voelz, who later became the director of the Los Alamos Health Division, became famous, or infamous, for his role in the Karen Silkwood incident. Silkwood was a chemical technician in the Kerr–McGee plutonium fuels production plant in Crescent, Oklahoma. One of her jobs was grinding and polishing plutonium pellets that were going to be used in reactor fuel rods. On the evening of November 5, 1974, she found, by measuring with a monitor, that the right sleeve and shoulder of her coveralls exhibited alpha-particle activity, which suggested the presence of plutonium. She went to the plant Health Physics Office and was given a nasal swipe. It showed a modest amount of activity. The gloves that she had been using were replaced and the old gloves analyzed. The first of the several anomalies that kept turning up in her case was that no leaks were found in the gloves and that the air monitors in the room where she had been working showed no activity. It was never clear where the plutonium came from. She was put on a program where urine and feces were to be collected for five days. She was assigned work that would not put her in contact with plutonium, and when she left the plant after her visit to the Health Physics Office, she again monitored herself and found no activity. But the next day, when she returned, the alpha activity was back. When, on November 7, she once again reported to the Health Physics Office, her bioassay samples showed extremely high levels of activity, even though she had had no contact with plutonium at the plant since November 5. Both her locker and her car showed no activity, but her apartment, which she shared with another Kerr–McGee plutonium worker, did. This was enough to have Kerr–McGee send Silkwood, her colleague, and Silkwood's boy friend, who had been spending time in her apartment, to Los Alamos for testing and hence to Dr. Voelz.

Dr. Voelz found that Silkwood was emitting gamma rays—electromagnetic radiation with a shorter wavelength than x-rays. The explanation was the following. In the reactor-produced plutonium, as we have seen, plutonium-240 is a by-product. But plutonium 240 can absorb yet another neutron, becoming plutonium-241. This isotope beta-decays into the next element in the periodic table, americium. It had been identified from its gamma ray emission by Seaborg and his collaborators at the Met Lab in 1944. From the intensity of the americium gamma emission, Voelz could estimate how much pluto-nium was in her body. He assured her that, based on his experience with other plutonium workers (including Magel, one supposes), the amount she had was not a danger either in producing cancer or in affecting her ability to have normal children. Whether he was right, we shall never know. While on the way back from a union meeting on the night of November 13, Silkwood was killed in a one-car accident. She was 28. An autopsy was performed that confirmed Dr. Voelz's plutonium estimates. However, a substantial amount of methaqualone (Quaalude) was found in her blood and in her stomach. Why she was taking this drug, usually prescribed as a sedative, I do not know, but she was taking enough so that it could have caused her to fall asleep. This, and other curious facts about her death, along with her activity in exposing what she perceived as serious flaws in Kerr–McGee's safety measures to protect the plutonium workers, inspired a conspiracy theory that still has its advocates. After her death, her estate filed a civil suit against Kerr–McGee. It went back and forth until it was finally settled out of court for $1.3 million in 1986. Kerr–McGee closed the plant in which Silkwood worked in 1975.

As one might imagine, Magel had a rather more nonchalant attitude toward these matters. In his 1995 interview, when asked if he was worried about his exposure to plutonium, he replied, "I didn't get too excited or worried about it. I am not super patriotic or anything like that, but it was war and we had a job to do." Then he went on to add:

By then, they knew from animal studies that plutonium goes to the bone [when breathed, ingested, or otherwise internalized]. They thought that if we built up our calcium content, there would be less reason for plutonium to want to reside there. They had to develop health procedures from scratch, because there was no plutonium before that time and, of course, no experience working with it. Nick [Dallas] and I were there, so we were guinea pigs for trying out new health procedures. We are also two of the original members of the UPPU club. We've been monitored for any damage that plutonium might cause. Every year, I would send them a gallon of urine from a 24-hour period so they could measure the plutonium content.

He continued:

I can't speak for all the UPPU members, but in 1971, they decided to bring all 26 of us back to Los Alamos to do complete physical exams and to get whole body counts, urine counts, x-rays, and blood work. They were using the urine data to measure the long-time excretion rate of plutonium compared to the amount retained. They're still collecting basic chemical and medical information on the rate at which the body rids itself of plutonium once there is an uptake.

They've also worked very hard to measure the amount in our lungs and to monitor our lung performance. They were looking for any effect that might confirm or dispute the news media claim that one speck of plutonium will kill the population of the Earth. The media keeps writing that story over and over to the point that I get very,

very, mad. I've been after George Voelz to write an article and stop this nonsense. Sure it is a hazardous material, but there are at least 26 of us who've been carrying it around for decades, and eighteen of us who, after fifty years, are still healthy and just getting older.

Voelz was himself interviewed at this time.[13] In neither of these interviews does he mention Magel or Silkwood by name. He does try to put the issues in perspective. I will quote what he said. How convincing one finds it can, I am sure, be debated. In the last chapter I present other perspectives. Here is Voelz:

> Let me begin with a few very simple facts. Each one of us in this room, without considering the effects of occupational exposures, has a one-in-three chance of getting cancer in our lifetimes. And we each have a one-in-five, or 20 percent, chance that we'll die from cancer. That means of the 21 people in this room, 7 of us will probably get cancer, and 4 of us will probably die of cancer.
>
> Now if your occupational exposure is within the limits set by the Department of Energy, or even if your exposure is well above those limits, your increased risk of getting cancer is not so very great compared to this basic rate. The problem is that if you get cancer you begin to wonder, "Did I get it from the radiation exposure?" And there is no way to answer that question because there's no way to tell whether radiation was the cause. As a physician responsible for the health of radiation workers, that bothers me a great deal.

Voelz goes on:

> Another thing that bothers me is our past failures in communication. . . . The medical people were doing a lot of worrying and studying and thinking behind the scenes, but we probably didn't share enough of our thinking with the workers who were getting exposed. We had a particularly hard time monitoring inhalation exposures, because once plutonium gets in the lung, it may be anywhere from 6 months to several years before any of the material migrates to other parts of the body and shows up in the urine. In some autopsies, we've

seen that 30 or 40 years after the exposure, 75 percent of the inhaled plutonium is still in the lung.

On the subject of communication, Voelz added, "I think we did much better with the members of the UPPU club. Those were the people who had unusually high exposures in the old D Building." This was the original building where the plutonium work had been done. It had what were then state-of-the-art ventilation systems. The original 26 were monitored from about 1948. As Voelz notes:

> The first official examinations were done by physicians in the areas where they were living in about 1952. It's been about 50 years since most of them had their major exposures in 1945, so this is a sort of golden anniversary for them . . . they've fared pretty well as a group. Of the original 26, only 7 have died, and the last death was in 1990. One was a lung-cancer death, and two died of other causes, but had lung cancer at the time of death. All three were heavy smokers. In fact 17 of the original 26 were smokers at the time they worked in D Building. Smoking was a very social activity during World War II. The military offered free cigarettes, and if you turned someone down when they offered you a cigarette, it was almost taken as an insult.

Oppenheimer was a chain smoker and ultimately died of throat cancer.

Voelz continues:

> In any case there were three deaths involving cancer, which is consistent with the national cancer mortality rate for a group of this size and age. Then there were three deaths due to heart disease and one due to a car accident. According to the national mortality rate, one would have expected 16 deaths in this group by this time, so the mortality rate for the group is about 50 percent lower than the national average. That's due to good lifestyle more than anything else. People who are well behaved, predictable, and responsible generally live longer than the average, and those are the characteristics selected for in plutonium workers.

I am not sure how Magel and Dallas fitted this profile.

We compared the mortality rate of the 26 UPPU Club members with the rate of unexposed Los Alamos workers from the same period. This comparison eliminates the so-called healthy-worker effect, the fact that the employed population has a lower frequency for disability and disease than does the general population. The risk ratio for all causes of death was 0.60 and from deaths from all cancers was 0.82. A risk ratio of less than 1.0 indicates that the risk of death in the exposed group is less than in the unexposed. Because of the small number of people in the exposed group, even these low ratios were not statistically significant. Nevertheless, it is of some considerable comfort that they are low.

We recently published a study of all the males who have been employed at the Los Alamos Laboratory during the period from 1943 through 1977. That is some 15,000 people. The important finding from the standpoint of radiation, is that we did not find any increase in the rate of leukemia or other blood-cell cancers that tend to increase with increasing exposure to radiation. We did a trend analysis that showed the rate of three cancers (esophagus, brain, and Hodgkin's disease) correlated statistically with increasing exposures to doses of external radiation. These particular cancers, however, have not been known to be caused by low-dose radiation in other studies. This inconsistency made us conclude that the significance of the observed findings was indeterminate. We also compared cancer rates in workers exposed to plutonium with those in unexposed workers. There were no statistically significant elevations of cancers in the plutonium-exposed workers.

Voelz concluded:

So far, we have not seen any significant health effects from pluto-nium, but that doesn't mean that plutonium isn't very hazardous. It is. But we have taken great care from the beginning to operate with conservative limits on the permissible body burden for plutonium workers, and those limits are not special for plutonium but rather are equivalent to the occupational limits placed on all types of radiation exposure.[14]

We return to these matters in the last chapter.

Essentially all of the plutonium used during the war was made in the reactors at the Hanford Engineering works in Hanford, in Washington State, which was built along the Columbia River. Figure 12 shows plutonium production at Hanford year by year. The distinction between reactor-grade and weapons-grade plutonium refers to the amount of isotopes, particularly plutonium-240, in the mixture. Weapons-grade plutonium can have no more that 7 percent. The fact that wartime plutonium was all weapons grade really had to do with the sense of urgency about production. Plutonium was taken out of the reactor as quickly as possible, after about a hundred days of irradiation. As will be discussed in Chapter XI, the names "reactor grade" and "weapons grade" are misleading. You can make a bomb out of reactor-grade plutonium, which is something to keep in mind when discussing proliferation. Note that the total production from 1945 to 1947 was 500 kilograms, a little more than half a ton.

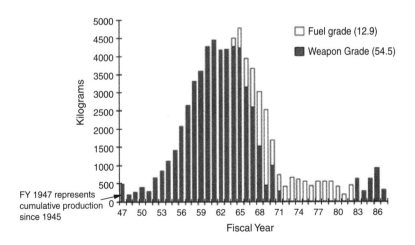

Figure 12 Hanford plutonium production (total of weapons grade and fuel grade).

Compare this to the micrograms with which Zachariasen worked and you will see what a difference these reactors made.

The first production reactor built at Hanford was the so-called B reactor shown in Plate 12. You can see the Columbia River in the background. The reason for the proximity to the river arose from a decision of the Princeton physicist Eugene Wigner, who was responsible for much of the conceptual design. Wigner, who won the Nobel Prize in physics in 1963, was one of the most mathematically sophisticated of twentieth century physicists. But he had a degree in engineering, and a good deal of the reactor theory one finds in books is due to him. There are three basic elements in a reactor: the fuel, the moderator, and a coolant. In the B reactor the fuel was natural uranium that had been processed in Canada. The moderator was highly purified graphite—the moderator material that had been used in Fermi's reactor. The role of the coolant is to prevent a meltdown in which the fuel elements overheat and melt. In the much smaller Clinton reactor, the coolant had been air. But Wigner decided that this would not work as readily with a large production reactor, so he chose water cooling. The water came from the Columbia River and was returned there.

The B reactor was first allowed to go critical on September 27, 1944, at 12:01 a.m. I know the time because John Wheeler, who was there, told me about it. He also told me that a number of dignitaries, including Fermi, were on hand. Wheeler was there because he had been given the job of theoretical physics consultant to the project. He explained that when he first got to the site, there was a kind of Wild West atmosphere. There were some 50,000 workers and a good deal of rowdy behavior.

Wheeler was in the central laboratory computing various things when the reactor went critical. Soon he began getting very disturbing reports. When the safety rods were pulled out so that neutron multiplication could rise, the reactivity also rose as it should have done. But then, for no apparent reason, the reactivity started to fall. Even with all the safety rods pulled out, as Wheeler put it, it "died to death."

"Everybody was scurrying around," Wheeler recalled, trying to figure out what had happened. "But," he explained, "it had been one of my jobs to consider every possible way that things might go wrong. I was, therefore, very aware that a fission product, when it decayed, could give rise to another one which could absorb the neutrons. When a few hours later, the reactivity began rising again, I became sure this was what had happened. Now the second nucleus had decayed into a third one which did not absorb neutrons." Wheeler came to the conclusion that the first fission fragment was iodine-135, which is produced in about 6 percent of the fissions. It beta-decays with a half-life of 6.6 hours into xenon-135. This isotope is much more absorbent of neutrons than even uranium-235. So long as it is present it disrupts the chain reaction cycle. But with a nine-hour half-life, it beta-decays into cesium-135, and this isotope does not absorb neutrons. The antidote to this poison is to increase neutron production so that the fission neutrons can overpower this effect. Wheeler explained what happened. "The hero of the story," he said, was a DuPont engineer named George Graves. He'd kept asking these questions like 'What in the hell are fission products?' Once he got into it, he insisted that instead of the 1500 fuel tubes we had planned, we have a margin for error of another 500—actually we had 2004. That decision took a lot of gumption, since it cost a lot of money. But thanks to his foresight it was possible to reload those extra tubes and give the pile the reactivity it needed to override the fission product poison."[15] Wheeler also told me something else that, until that time, I had never heard. The Japanese sent paper balloons with incendiary bombs across the Pacific. Some of the balloons set fires in the Pacific Northwest, but one draped itself around the power lines that fed the water pumps to the Hanford reactor and shut it down. This was an incident that was kept completely secret during the war. In any event, it did not shut down the reactor long enough to stop production of the plutonium that was used to destroy Nagasaki.

The first plutonium from Hanford arrived in Los Alamos in February 1945. Thereafter, Hanford became the only wartime source of plutonium. Now the metallurgical problems became crucial: how to use plutonium metal despite the complexity of the allotropic phases. Smith had spent much of his professional life working on brass, which is an alloy mainly of copper and zinc in various proportions, depending on what it is being used for. It was natural for Smith to think in terms of alloys when it came to plutonium. He had been thinking about them before the laboratory began receiving large quantities of plutonium. Now the matter was urgent. There does not seem to have been much, if any, theory behind what was tried. To this day, as far as I can tell, there is still no theory on which everyone agrees. There is no theory that tells you what element to alloy with plutonium so that, for example, the δ-phase becomes stable at room temperature or indeed, once having produced stability, how long it will last. Finding suitable elements was a matter of trial and error. The first element that seemed to work was aluminum. A few percent of aluminum, when alloyed with plutonium, produced a metal in the δ-phase that was stable at room temperature. The problem was that aluminum, when impacted with alpha particles from plutonium decay, produced neutrons that would have complicated the design. But then it was discovered that gallium, when alloyed, also stabilized the δ-phase. Aluminum has an atomic number of 13, while gallium has an atomic number of 31. The extra positive charges in the gallium nucleus repel the positively charged alpha particles and inhibit neutron production. Thus, gallium seemed ideal. The difficulty was that because of the extreme time pressure under which the laboratory was working—it was now the spring of 1945 and the bomb was scheduled to be tested early that summer—there was no time to study how long such an alloy would maintain the stability of the δ-phase. The last thing one wanted was for it to revert prior to the explosion to the α-phase, with all its attendant difficulties. In his short memoir, Smith recalls that the metallurgist Eric Jette, who had actually worked with the alloy, strongly opposed using it without a stability test for which there was no time. Smith's instinct told him that it would work, but the decision to use it was not one he could make on his own. He felt he had to consult Oppenheimer.

The two of them had dinner at Edith Warner's "house at Otowi crossing." Edith Warner was a Pennsylvania woman who had come, when she was in her mid-thirties, to New Mexico for her health. She lived in a house at the Otowi railroad crossing with the governor of the San Ildefonso Pueblo, an American Indian named Atilano "Tilano" Montoya. The house had originally been the post office and supply storage facility for the Otowi railroad station. Oppenheimer, who had also originally come to New Mexico for his health, had gotten to know her before the war. She had opened a tea shop at her place where she served chocolate cake to, among others, the boys from the Los Alamos Ranch School. The school was later taken over when the site was chosen for the laboratory. During the war, no doubt with encouragement from Oppenheimer, she also served dinners to a small and select clientele from the laboratory. It was over such a dinner that Oppenheimer listened to Smith's concerns and told him to make any decision he thought was right. Smith opted for gallium, and the weapon that was tested at Alamogordo, and then dropped on Nagasaki, had 0.8 percent by weight, or 3 percent by molecular content, of gallium. One very advantageous feature of this arrangement was that at relatively low pressures this alloy reverts to the α-phase. This, as I have mentioned, has a substantially higher density and therefore lower critical mass. It turned out that implosion, to which I turn next, produced pressures sufficient to provoke this phase change, thus increasing the efficiency of the bomb.

The first suggestion that implosion might be used to obtain a critical or supercritical mass, at least by the Americans, was made in the summer of 1942. At a secret meeting at Berkeley, people such as Oppenheimer and Serber discussed the state of bomb physics. The Caltech physicist Richard Tolman proposed using implosion, and this idea found its way into Serber's primer with the diagram in Figure 13. It is a pretty impressionistic drawing, but the idea was to distribute explosives around the ring, which would be made of, say,

FIGURE 13 Implosion from Serber's primer.

uranium-235. The explosives were to blow the pieces of uranium inward so that they would make a sphere that would be critical or supercritical. This is not the way the actual bomb was being designed, but Serber relates that Oppenheimer fielded some questions about it from the powers in charge at the time. He told them that Serber was working on it, which was the last thing Serber was doing. On the other hand, after Serber gave his lectures in the spring of 1943, a physicist named Seth Neddermeyer took an interest and headed a small group that looked into some aspects of implosion before it became necessary. When, in the summer of 1944, the laboratory was almost totally converted to the study of implosion, Neddermeyer must have thought he was being drowned by a tidal wave. Two people who were not involved were Serber and Edward Teller. Oppenheimer told Serber that he should take charge of the uranium bomb enterprise, which was still active, and Teller wanted to work only on the hydrogen bomb.

Implosion is really not a plutonium story. What motivated it, as I have said, was the discovery of plutonium-240 in the reactor plutonium. What was involved was how to use high explosives to rapidly compress a plutonium sphere. At no time before the Alamogordo test were spheres of actual plutonium used in any of the experiments. Aluminum and natural uranium were used. At first, an attempt was made to implode a more or less hollow sphere. The results were problematic. At the end of 1944, the laboratory switched to the study of what became known as the "Christy gadget." Robert Christy had been a student of Oppenheimer's. He was doing some theoretical studies on implosion that led him to believe that a solid, or nearly solid, sphere would implode more uniformly. It was decided to try this approach. The sphere was to be manufactured from two hemispheres, each of which would be coated with nickel to keep the plutonium from oxidizing. What happened when the Los Alamos scientists attempted this for the Alamogordo Trinity test bomb was described by Smith. He wrote:

> . . . the hemispheres for the . . . Trinity test were electro-plated and some aqueous electrolyte retained in a porous spot in one of them reacted and caused a tiny blister to form. This would have separated the mating surfaces enough to allow jetting during implosion and possible premature initiation. Postponement of the test was threatened, but I proposed the insertion of some rings of crinkled gold foil to prevent jetting and, late one night, I had the by-then-rare experience of working in the laboratory to make something with my own hands instead of watching someone else do it following instructions. This little blister made it necessary for me to become at the last moment a member of the team that was responsible for the final assembly of the first nuclear bomb. At approximately noon on 15 July 1945, at MacDonald's Ranch near Alamogordo in New Mexico, I put the proper amount of gold foil between the two hemispheres of plutonium. My fingers were the last to touch those portentous bits of warm metal. The feeling remains with me to this day, thirty-six years later.[16]

After the war, Smith couldn't face the prospect of returning to the Naugatuck Valley and American Brass. He went first to the University of Chicago and then finished his career at MIT. While there, he wrote books and articles on the connection between the history of the decorative arts and the development of metallurgy. He died in Cambridge, Massachusetts, on August 25, 1992, at the age of 88.

In the next relatively brief and somewhat more challenging chapter, I am going to try to explain why plutonium is as bizarre as it is.

X
Electrons

The radius of Np^{+4} is thus 0.015 Å larger [The symbol Å stands for angstrom and is 10^{-8} centimeter. The superscript +4 indicates that we are dealing with an atom that has given up four electrons and is now a positively charged ion. Zachariasen is talking of differences in radii here. Later the modern values of the radii themselves are given.] than that of Pu^{+4}, 0.016 Å smaller than that of U^{+4}, and nearly identical with that of Ce^{+4}.

I believe that a new set of "rare earth" elements has made its appearance. I believe that the persistent valence is four, so that thorium is to be regarded as the prototype just as lanthanum is the prototype of the regular rare earth elements.

W. H. Zachariasen, June 1944[1]

This chapter is the most technically demanding chapter in the book. In it I am going to try to explain the very strange physics and chemistry of plutonium. Even if plutonium behaved in a less anomalous way, such an explanation would make demands on the reader because it inevitably involves quantum mechanics, which itself is not easy to explain. However, with plutonium we have the added complication of its oddity. Because of these difficulties, I have decided to present a kind of "summary" of what is involved that glosses over the details. Some readers may find that this is all they want to know. They are welcome to skip the body of the chapter where these details are discussed. The next, and final, chapter concerns the politics of plutonium and can be read without knowing these details.

Any discussion of the science of plutonium must be divided into two parts: discussion of the individual plutonium atom and discussion of ensembles of these atoms as they would be manifest in the crystal structure of metals. The physics in the latter case reflects the collective behavior of these atoms. We begin with the individual plutonium atom. Like all atoms, it has a positively charged nucleus surrounded by negatively charged electrons whose total charge just balances the positive charge of the nucleus. Because these electrons are described by quantum mechanics, their properties are limited. For example, they are restricted in both the energies they can have and their angular momenta. In building up the atom conceptually, imagine putting the electrons in one at a time and filling successive "shells" of electrons following the laws of quantum mechanics, which among other things restricts the number of electrons in each shell. When a shell is filled, the succeeding electrons go into new shells. If all of the allowed shells are then exactly filled, the resulting atom—neon is an example—will be chemically inert. If the last shell is only partially filled, these electrons—so-called valence electrons—can take part in the chemistry. The electrons can join other atoms that have partially filled shells to produce a chemical bond. The naïve expectation is that when you go from one atom to the next in the periodic table you add an electron, which would change the chemistry. When the transuranics were first discovered, it was learned that this is not what happened. The chemistries of uranium, neptunium, and plutonium, for example, are very similar. This means that the added electrons are not taking part in the chemistry and are shielded from the valence electrons that do, which remain the same in these elements. The body of this chapter contains more details, but this is an important property of plutonium that must be accounted for.

Atoms are not like golf balls or marbles. They do not have definite shapes and sizes. The reason again is quantum mechanics. Electrons in an atom do not have definite positions, only probable positions determined by quantum theory. This means that there is no unambiguous way of defining the "size" of an atom. The body of this chapter presents three different ways of defining the size, which lead to different ways of measuring it, with different answers. Therefore, when we discuss the size of a plutonium or neptunium atom it is important to specify what we mean. Even so, these different sizes do have some commonalities. There is one in particular to which I would like to call your attention. It is illustrated later in Figure 15, which plots the so-called ionic atomic radii for the lanthanide and actinide series of elements. In the body of the chapter I explain what is meant by the ionic radius of an atom. The actinide series includes the transuranics. The lanthanides are often referred to as "rare earths." The striking thing about Figure 15 is that the radii shrink as the elements get heavier. Heavier elements have more protons in their nuclei and hence more electrons in their shells. The added protons produce an added positive charge, which increases the attractive force on the electrons. The shells are pulled closer to the nucleus—hence, the contraction. That this contraction exists for the transuranics was discovered early in the war. This is summary of what I want to say about individual atoms.

In collections of atoms, new quantum mechanical effects appear. Let us begin with the case of two sodium atoms. If these atoms are widely separated, then what matters are the properties of the individual atom. Sodium is an atom with all its shells filled except one. In this last shell there is a single valence electron responsible for its chemistry. The minimum energy that this electron can have is called its "ground-state" energy. If the atoms are separated, each of the valence electrons has this energy in its ground state. But if the atoms are in close proximity, as they would be in a crystal lattice, then the probability distributions in space of these electrons overlap and the allowed energies are different, reflecting this overlap. One possible allowed energy for the isolated atom becomes two distinct allowed energies for overlapping atoms. If instead of two atoms we have 10^{22} atoms per cubic centimeter, as we would in a sodium metal, there are so many possible energies that, to all intents and purposes, they form a continuum. An electron can wander around in this structure more or less freely. It is no longer attached to an individual nucleus but becomes what is known as "itinerant." In a metal, these itinerant electrons, which are shared among the atoms in the lattice, contribute to the binding. The more there are, the tighter we expect the metal to be bound. This would suggest that as you move across the actinide series there would be more electrons, hence more itinerant electrons, and thus tighter binding and a smaller radius of the crystal. As Figure 16 shows, this is the case until we come to plutonium. At plutonium this radius increases, suggesting that the number of itinerant electrons has decreased. In fact, it appears as if these plutonium electrons can't decide whether they are bound or not. The details of this curious situation are the subject of much current research. There seem to be promising models but none that are universally accepted. A few more details are given in the body of this chapter.

This closes the summary of the chapter.

In the late 1950s, when I was a visiting member at the Institute for Advanced Study in Princeton, Oppenheimer, our director, had a mantra that he would trot out when asked to describe how the physicists actually went about their work at the institute. "What we don't understand," he would remark, "we explain to each other." In writing this chapter I have often invoked a version that goes, "What I don't understand, I explain to myself." Part of what I don't understand, no one seems to understand. The scientific study of plutonium is a work in progress. It is a very active subject, even for many people who have no association with nuclear weapons. In studying some of this literature, I find that it has all the characteristics of debate and controversy that any live and interesting scientific subject has. But, here, there are some applications—for example, the long-run safety of the plutonium-based nuclear weapons we have in storage—that make it rather different than, say, the study of string theory. However, the unfinished status of plutonium science is not the only, or even the most important, reason for my reticence. My field of physics, cosmology and elementary particles, is very far removed from the physics that is used in these studies. In this part of physics I am an amateur, and like many amateurs, there is—in Bohr's wonderful admonition—the temptation to speak more clearly than I think. With this caveat I can tell you what I have learned.

There are at least two domains in which the behavior of plutonium is strange. Individual atoms have properties that were surprising and unexpected, and the atoms, behaving collectively, as in plutonium metal, also have properties that were surprising and unexpected. The allotropic phases are an example. In both of these domains it is the atomic electrons that account for what is observed, but quite differently. Electrons in individual atoms interact with the positively charged nucleus and with each other, while electrons in, say, a plutonium metal interact with a variety of sites—charged plutonium ions—in the metal. Indeed, under these circumstances, it is closer to the truth not to attach some of the electrons to any particular site at all. I am going to begin my discussion with the case of the isolated plutonium atom because I think that the issues are simpler and because much can be revealed. I start with atomic sizes, which is what the quote from Zachariasen above refers to. At once, we have a problem.

Unlike billiard balls, or marbles, individual atoms do not have sizes in the usual sense. To see what is at issue, consider hydrogen. The hydrogen atom consists of an electron and a proton nucleus. The electron can be found anywhere. Quantum mechanics tells us how probable it is to find the electron at some particular distance from the proton. With exponentially decreasing probability, the electron can be found as far away as you like from the proton. Given the fuzziness of the shape, chemists measure—or define—atomic size in various ways. Here are three definitions of an atomic radius. The first is what is called the "covalent" radius. Covalent compounds are compounds in which the elements involved share their electrons. A canonical example is the hydrogen molecule, which consists of two hydrogen atoms, each of which has a single "valence electron." When these atoms are close together, the electrons are shared between them, which is what produces the binding. The covalent radius is defined as half the distance between the nuclei of the two elements being held together by the covalent bond. Figure 14a gives the general idea. The covalent radius would be r. But many compounds are bound when one atom loses its electrons to another. This is called "ionic bonding." The canonical example is the compounding of sodium and chlorine to make salt. The sodium atom has one electron in its outside, or valence, shell, while the chlorine atom has a "hole" in its valence shell. The chlorine takes one of sodium's electrons, giving the chlorine atom a negative charge while the sodium atom acquires a positive charge. The ionic radius is determined from the distance between adjacent nuclei, but as Figure 14b shows, there is a problem.

In general, the two ions have very different radii. So-called cations (pronounced *cat-eye-ons*), which are formed when an atom loses electrons, have smaller radii than their parent atoms, while anions (pronounced *an-eye-ons*) formed when an atom receives electrons, have larger radii than their parents. The method described gives the sum of these radii. To find the individual radii, chemists have to resort to tricks. For example, the oxygen molecule has two identical oxygen atoms, so you can find the radius of an oxygen atom

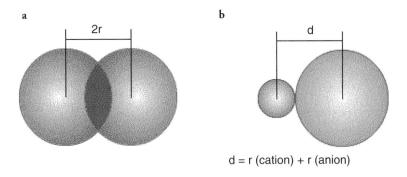

d = r (cation) + r (anion)

FIGURE 14 $2r$ and $d = r$ (cation) + r (anion) diagrams.

by dividing the distance between the oxygen nuclei in the molecule by 2. Then you can try to ionically bond the atom, whose radius we want to know, to oxygen and subtract the oxygen radius. A third definition of atomic size is the so-called metallic radius, which is defined as half the distance between neighboring atoms in a metal or, more generally, in any crystal structure. A version of the crystal structure radius is known as the Wigner–Seitz radius. What you measure is the volume of a unit cell. Then you assume that these volumes are approximately spherical so that the volume of a unit cell is given approximately by $4/3\pi R^3$, where R is the Wigner–Seitz radius. The measured radius for a given atom evidently depends on which definition is being used. This must be spelled out in each case because the values can be very different. In general, the covalent radius will be smaller than the metallic radius because covalent binding pulls the molecule together. The metallic radii depend on the forces that hold the crystal together, which, as we shall see, can vary a great deal from atom to atom.

Although it is true that the quantum mechanical atom has no size in the usual sense, nonetheless quantum theory can be used to calculate the most probable value of the distance, for example, of one of the valence electrons from the nucleus. The distances calculated this way share the same sort of general features as the distances found by chemists using the definitions given above.[2] The simplest example is hydrogen, which has, in its state of lowest energy, one electron with no orbital angular momentum—a so-called s state. The energies of such electrons are characterized by a positive integer n. The hydrogen ground state can then be characterized as a $1s^1$ state, where the superscript 1 refers to the number of electrons with these quantum numbers, s refers to the orbital angular momentum, and 1 is the energy quantum number. If you carry out the calculation for the most probable distance of this electron from the nucleus, you will find a value known as the Bohr radius, since Neils Bohr first found it using a mixture of classical and quantum physics. Numerically it is 0.53×10^{-8} centimeter. It sets the scale of atomic sizes.

I want to skip over helium, the next element in the periodic table, and go straight to the third element, lithium, where there are important lessons to be learned. In the electron structure of lithium, there is a single valence electron outside a closed shell of two 1s electrons. This electron is bound to the atom with a so-called binding energy, the energy that would have to be supplied to extract the valence electron. If we denote the number of protons in the nucleus by Z, then in the case of lithium, $Z = 3$. The valence electron, in this case, is in an s-orbital angular momentum state, with an energy quantum number $n = 2$. The $n = 1$ states are filled. There are, according to the Pauli principle, only two allowed possibilities for the zero angular momentum case, one electron with spin up and the other with spin down. These two electrons fill the first shell. A third electron must be put into a different quantum state, hence $n = 2$. We may compute the binding energy of the negatively charged valence electron by assuming that it is bound to the nucleus by the three positively charged protons, ignoring for the moment the effect of the other two electrons. With this assumption, the ratio of the binding energy of the lithium valence electron to the 1s hydrogen electron turns out to be $(3)^2/(2)^2$ or 9/4. For a general Z and n, the ratio of binding energies is given, ignoring the other electrons, by Z^2/n^2. However, if we compare this theoretical prediction to the measured binding energy, there is no agreement at all. What has gone wrong?

The electron structure of lithium might suggest an answer. The two 1s electrons are closer to the nucleus than the valence electron. They shield the valence electron from the proton charge. Since there are two of them, we might be tempted to conclude that the effective charge, which we will call Z_{eff}, would be 1.0, but this is not quite right. The 2s electron can penetrate the shell of the 1s electrons and feel the full force of the three proton charges some of the time. This produces a modified Z_{eff} that we can determine from the actual binding energy. For lithium, Z_{eff} turns out to be 1.26. We can use the same calculational model to find the most probable distance of the valence electron in terms of Z_{eff}. In units of the Bohr radius, it is given in terms of n and Z_{eff}, by n^2/Z_{eff}. For lithium, the result is 3.2 Bohr radii. One can carry out this kind of calculation for a variety of elements. If you do this, some very interesting regularities appear. If you go across a row of the periodic table from left to right, doing the calculations you will find that the radii of the outermost valence electrons decrease.[3] On the other hand, if we do the same for the columns, the effect is just the opposite: The radii increase.

The regularities predicted by the simple quantum mechanical model are by and large borne out, which tells us that the quantum mechanical calculation of the average distance to the outermost valence electron gives us useful insights. For example, we can get a feeling for why the systematic properties of the radii are what they are. Very qualitatively, the radii get larger when we go down the columns of the periodic table because, in adding electrons, the valence electrons are progressively farther away from the nuclei and thus less attracted to its positive charge. On the other hand, when we go across the rows of the periodic table the effective charges due to the protons get larger, so the electrons are more attracted to the nucleus and are therefore closer. There are complications and, as the empirical values show, exceptions, but this is the general idea.

When we come to the heavier elements, with their dozens of electrons, we can well imagine the phrase that the philosopher William James applied to the inner world of infants—"one great blooming, buzzing confusion"—might well apply here as well. But when it comes to atoms, we are saved by the Pauli exclusion principle. It is what makes the electron shell structure periodic. Without this periodicity we would indeed have a great blooming, buzzing confusion. How does this work? As I have already noted, the Pauli principle states that no two electrons can be in exactly the same quantum state. But what, in this instance, characterizes a quantum state? In these considerations of the periodic table, a quantum state is characterized by three attributes: its energy, its orbital angular moment, and the direction of its spin. The energy, in turn, is characterized by a positive integer, n, as in the 1s or 2s states of lithium. Here, 1 and 2 are the respective values of n. The orbital angular momentum, on the other hand, is more complex. In quantum theory it too is characterized by an integer, which by convention is denoted by l. For what follows, I need four examples starting with $l = 0$. The $l = 0$ state, from an ancient spectroscopic convention, is designated by the letter "s." We have already used this notation in discussing lithium. The three others we need are as follows: $l = 1$ is the p state; $l = 2$ is the d state; and $l = 3$ is the f state. These letters, too, came out of the early spectroscopic tradition. In classical physics, the angular momentum can have any value you like and the vector that describes it can point in any direction you like. In the quantum world, both the values and the directions are limited—"quantized." Quantum theory teaches us that a state with an orbital angular momentum l has $2l + 1$ substates that, very loosely speaking, correspond to the allowed directions in which the angular momentum vector can point. The spin, on the other hand, can point in only two directions, which we can designate "up" or "down."

Now we can see why the Pauli principle makes the periodic table "periodic." Take the first full row, starting with lithium, whose electron structure we have already discussed. We can move across the row, adding one electron at a time. The next element in the row after lithium is beryllium. It has two $n = 2$ electrons, one with its spin up and the other with its spin down. Both electrons can be added, with each having no orbital angular momentum—s electrons. However, to go to the next element in the row, we must add an electron with a different angular momentum so as not to violate the Pauli principle. The simplest thing to do is to add a p-state electron with the same $n = 2$ energy. We may ask, How many more such electrons can we add with s and p states before we run out of possibilities for elements in this row? What is the maximum number of elements in the row? If we count the s electrons, including the spins, we have two possibilities. For the p electrons, $2l + 1 = 3$, and if we add in the spins we have six possibilities, which brings us to a total of eight. We can't add any more electrons without going to the next row. Adding eight electrons lands us at neon. The neon electrons fill all the shells. This is why neon is chemically inert. Thus, the first period in the periodic table consists of eight elements, beginning with lithium and ending with neon.

The thing I want to call to your attention is a nice shorthand that chemists have developed for describing electron configurations. They realized that there is no point in repeating the configuration of neon, which is common to all of the elements in the row so they simply write—for example, for sodium—the configuration $[Ne]3s^1$, which means that there is, in addition to the neon shell, one $n = 3s$ valence electron. These outside valence electrons are what determine the chemistry of the element.

The reader will be relieved to know that I have no intention of going through the entire periodic table like this. My goal is to discuss plutonium, but a certain amount of background informa-tion is necessary. In particular, we must turn next to the so-called lanthanides. In a modern periodic table they occupy the row just above the row that includes plutonium—the actinides. If you do a web search, you will find that some sites begin the lanthanides with lanthanum. I think that the issues I want to discuss are best empha-sized, as in Figure 15, by beginning with cerium (Ce) and ending

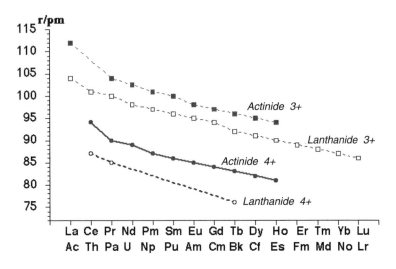

Ionic radii of two classes of ions: lanthanides and actinides.

with lutetium (Lu). These 14 elements have many commonalities. They are silvery white metals that tarnish when exposed to oxygen. In general, they have similar chemical properties. By 1940, they were found to have another property that was first recognized, and named, by Zachariasen's teacher Viktor Goldschmidt. It is what Goldschmidt referred to as the "lanthanide contraction." What this means is clear from Figure 15. The figure shows two curves of ionic radii for ions that have gained three or four electrons, respectively. You will notice that as we go across the row the ionic radii get smaller. This is the lanthanide contraction. The figure shows a similar phenomenon for the actinides but with a difference that we come to shortly.

Figure 15 shows the ionic radii for two classes of ions with three and four positive charges, respectively, for both the lanthanides and the actinides. The distance units are in picometers (10^{-10} centimeter). The contraction is evident for both types of elements. From what we have already discussed, we know what must be going on. The number of positively charged protons increases as we go across the row. For them to be effective in drawing in the electrons, the shielding by the electrons must be less effective. This gives us a clue.

Looking at the periodic table on page 13 we see that the element that precedes the lanthanides, and has all its electron shells complete, is xenon (Xe). This is another chemically inert noble gas. Thus the lanthanides have the xenon electron structure as a core, outside of which are the valence electrons. But we know from experiment that all the lanthanides have sensibly the same chemical reactions. This tells us that when we add an electron to go from one element to the next across the row, it cannot be one of the valence electrons that take part in chemical reactions. These valence electrons must remain the same across the entire row. By 1940, it was understood that what was happening was that the 4f shell was getting filled. These electrons have angular momentum 3 and energy quantum number 4. By the quantum rule presented above, angular momentum 3 electrons can be in $2 \times 3 + 1 = 7$ different states. Each one can have spin up or down, making 14 distinct states in all, which accounts for the 14 elements, from cerium to lutetium, that make up the lanthanide row. In orbits outside the 4f electrons, the s and d electrons take part in chemical reactions. For example, the configuration for cerium is $[Xe]4f^1 5d^1 6s^2$, which means that outside the xenon electron core there are one 4f electron, one 5d electron, and two 6s electrons. If we go across the row to lutetium, the configuration is $[Xe]4f^{14} 5d^1 6s^2$, which means that the f shell has filled up with 14 electrons, but the valence electrons responsible for the chemistry have remained the same. The reason for this is that the high angular momentum of the f-shell electrons keeps them for a considerable fraction of the time outside the valence electrons that are responsible for the chemistry. These electrons have a lower angular momentum and when they are inside the f electrons they cannot take part in the chemistry, Thus, the lanthanides all have substantially the same chemistry. The added proton charges pull in the valence electrons, which is what accounts for the observed contraction of the ionic radii.

All of this seems to have been well understood by 1940. It even appears in standard textbooks of the period.[4] It was a premise of the paper that Maria Mayer published in 1941. By 1944, as the quotation from Zachariasen at the beginning of this chapter makes clear, the same set of phenomena was once again appearing with the set of elements beginning with thorium and, as far as Zachariasen could verify, persisting through plutonium. Once again, the chemistry is sensibly the same for these elements, and the ionic radii appear to be contracting. Given all this, it is puzzling to me that in 1944, when Seaborg suggested what seems like the obvious explanation, he was, according to his account, treated as if he had just lost his mind. But certainly once the data were declassified, it was generally agreed that this series—the "actinides"—had, outside a radon core, an increasing series of 5f electrons, with the valence electrons that take part in the chemistry outside them. With plutonium, typically, there is a complication in a way that is a little surprising. It involves Einstein's theory of relativity.

In these very heavy atoms the electrons move at speeds close to that of light. This means that effectively they are more massive, which changes the probability distributions. There is some probability that the 5f electrons are found at larger distances from the nucleus than might have been expected. This means that unlike the lanthanides, these electrons can play a role in chemical reactions. The 5f electrons in the actinide series start at protactinium, which is actually the second actinide after thorium. Thorium does not have a 5f electron. The electron structure of protactinium can be written symbolically as $[Rn]5f^26d^17s^2$, where Rn stands for radon. All of the actinides have a common radon core. Radon is another noble gas—element 86. To go from one member of the series to the next, one adds new 5f electrons. Uranium has three and neptunium four. Plutonium has six. The series ends with lawrencium, which was first made in Berkeley in 1961, whose configuration is $[Rn]5f^{14}6d^17s^2$. The 14 5f electrons fill the shell.

In fact, until the mid-1960s and early 1970s, it was assumed that the presence of these atomic 5f electrons explained everything about plutonium, including its bizarre properties as a metal. In a sense this is true, but not in the way those early physicists thought. Once again we need some background. Let us start with sodium metal as an illustration. Atomic sodium has one s electron outside a neon shell. This electron is normally in its state of lowest energy—the ground state. If there are two sodium atoms separated from each other, then each associated electron will have the same ground-state energy. But if the two atoms are brought close together, the probability distributions associated with these electrons will overlap. When this happens, neither electron belongs to one atom or the other. In a sense, each electron belongs to both. The energy that such an electron can have is not the same as the energy the individual electrons had. There are now two new energies possible for the electrons. It is the overlapping of the electrons that is responsible for binding the two sodium atoms together to make a molecule. We can now imagine making a lattice of closely spaced sodium atoms, which is what metallic sodium is. Instead of two atoms, we have more like 10^{22} atoms per cubic centimeter. There are so many possible energies that, to all intents and purposes, they form a continuum—a band—between a maximum and a minimum energy. An electron that finds itself in this band can wander freely throughout the metal. This is what the practitioners in this field refer to as an itinerant electron. It is these itinerant electrons that bind the metal together.

The picture that has emerged for plutonium, and the other actinides, is somewhat similar to this one. The details are complex, and under active study, but here is the general idea. As I have noted, the actinides develop along the row by adding 5f electrons. In a metal, these electrons either can be localized to the atomic sites or can be itinerant or sometimes neither. Which occurs, it turns out, depends on the actinide. There is a point about the localized electrons that is interesting, which I have mentioned: They move around their atomic nuclei at speeds that are close to that of light. This means that Einstein's theory of relativity has to be used to describe them correctly. One usually does not think of using the theory of relativity for something as staid as a metal, but there it is. Both theory and experiment suggest the following picture.

For actinides that are less massive than plutonium, the 5f electrons are itinerant. This means that they contribute to the binding of the metal. As you go across the row, you would expect the metallic radii to decrease since you are adding itinerant electrons. This is what is observed. Figure 16 shows the Wigner–Seitz radii, which are approximately the metallic radii. Note that up to plutonium there is a sharp decrease in the radii, which then continues more slowly after americium. The actinides up to plutonium behave like the transition metals and after americium like the lanthanides. Plutonium marks the transition. It is at plutonium that the radius jumps up, and indeed, the δ-phase radius seems to have a mind of its own. What seems to be happening is that at plutonium some of the 5f electrons

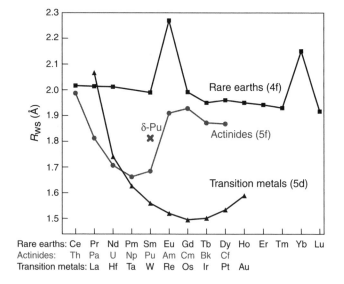

FIGURE 16 The Wigner–Seitz radii in angstroms for three classes of atoms. Note the striking behavior for the actinides at plutonium.

become localized. When this occurs they no longer contribute to the binding and the unit cell volume increases. From americium onward all of the 5f electrons are localized and the radii decrease slowly, presumably because the number of protons is increasing as you go across the row. The weird behavior of plutonium is attributed to the fact that its electrons can't decide whether they are bound or itinerant. For example, the δ-phase electrons are somewhere in between. Hence its unit cells have a volume that is off the curve. This knife-edge behavior of plutonium leads to its instability as shown by its six allotropes that can change phase with small perturbations of the external pressure and temperature. It would be nice if there were a simple explanation for these things, something like explaining why a rock falls when you drop it, but this is in the quantum world.

I want to end this chapter by describing a bit of plutonium chemistry that has implications for the present and future status of plutonium, which is the subject of the next, and last, chapter. If you expose plutonium metal to air, it oxidizes. A surface layer of plutonium oxide is formed that tarnishes the metal. But if you expose the metal to more oxygen, in the presence of water vapor, the oxidation rate is increased by very large factors. This action produces heat and the plutonium can begin to burn—indeed it can produce a dangerous fire. This is especially true if the plutonium metal is in powder form. In this form it can ignite at the very low temperature of 150°C. In 1957, and again in 1969, there were two fires at the Rocky Flats Plant, located in Golden, Colorado, near Denver.[5] In the mid-1950s the plant, which had begun production of plutonium pits in 1952, began making hollow pits. There was great pressure to turn out as many of these pits as possible in the shortest time. This haste led to both fires. At the site of the 1969 fire there were plutonium briquettes that had been made from plutonium scrap metal. They were stored near some oily rags that were also contaminated with plutonium. The water vapor in the air ignited the plutonium in the rags. This, in turn, caused one of the briquettes to ignite. The resulting fire caused tens of millions of dollars in damage. If it had not been for the heroic actions of the firemen, who repeatedly entered the building while the plutonium was burning, there could have been an environmental disaster with enormous consequences. This is an example of the postwar problems that have been, and are still being, caused by the production of plutonium. In the final chapter we discuss others.

XI
Now What?

For more than half a century, Oak Ridge National Laboratory has had a program of selling radioactive isotopes for research and medical uses. Its website, *www.isotopes@ornl.gov*, provides both a menu of the available isotopes and a description of the program. To find the prices, which change over time, you need to call (Figure 17). I decided to find the price of plutonium-239. Remember that the first half-gram lots of plutonium came to Los Alamos from the Clinton reactor in Oak Ridge in the spring of 1944. The cost in milligrams must have come to millions. I spoke to a very pleasant man at Oak Ridge who had the price list. He told me that the going price was $5.24 per milligram. I did not ask about uranium-235. I thought that two inquiries about potential bomb-making material might arouse suspicions, but a few years ago a colleague did ask and was told that it was $57 a gram.[1] I am not sure in what form the uranium is delivered, but at that price, it would cost about $2.4 million to buy enough to make a gun-assembly bomb. My guess is that if you tried to order that quantity you would get a visit from your friendly neighborhood agent of the Federal Bureau of Investigation (FBI). The bare critical mass of a δ-phase plutonium bomb is about 15 kilograms. In this context, what "bare" means is that you don't try to improve the bomb by, for example, adding a

Uranium-235

ISOTOPE °° ^{235}U	
Half Life / Daughter	7.04 x 10^8 years to thorium-231
Major Radiation	Alpha - 4.39 MeV
Form	Oxide
Activity	~ 2.16 uCi/g
Radiopurity	> 98%
PRODUCTION	
Source	Natural uranium
Processing	Electromagnetic separation of natural uranium
DISTRIBUTION	
Shipment	Glass bottle in a nonreturnable or returnable container
Availability	Stock. Classed as a nuclear material requiring documentation of transfer (DOE/NRC Form 741)
Unit of Sale	Milligrams
Note: Quantity discounts may be available. Call for current discounted price.	
Contact	Oak Ridge National Laboratory

FIGURE 17 Ordering information from Oak Ridge: uranium-235 and plutonium-239.

Plutonium-239

ISOTOPE ∘∘ ^{239}Pu	
Half Life / Daughter	24,100 years to uranium-235
Major Radiation	Alpha - 5.15 MeV
Form	Oxide powder
Activity	~ 61.3 mCi/g (theoretical)
Radiopurity	99.00-99.99%
PRODUCTION	
Source	Neutron irradiation of uranium-238 and electromagnetic separation of plutonium isotopes
Processing	Transuranic processing
DISTRIBUTION	
Shipment	Nonreturnable / returnable container
Availability	Stock. Classed as a nuclear material requiring documentation of transfer (DOE/NRC Form 741)
Unit of Sale	Milligrams
Note: Quantity discounts may be available. Call for current discounted price.	
Contact	Oak Ridge National Laboratory Isotope Business Office Call: (865) 574-6984 Fax: (865) 574-6986 email: isotopes@ornl.gov

FIGURE 17 Continued.

uranium tamper, which produces additional neutrons that can cause fission and slows down the expansion of the exploding material. What stops the explosion in an atomic bomb is the expansion of the material, which causes it to become sufficiently dilute so that the mean free path for fission is too long to sustain the chain reaction. That happens long before the uranium or plutonium has fissioned in its entirety. Indeed, in the Hiroshima bomb something like 98 percent of the explosive material did not fission, while in the Nagasaki bomb about 80 percent did not. Incidentally, by counting all the above-ground explosions of nuclear weapons, it has been estimated that something like 10,000 kilograms of plutonium was released into the atmosphere. Using the quoted price for plutonium-239, it would cost about $150 million to buy a bare critical mass worth. Incidentally, I was told that the transportation cost was in the thousands of dollars per milligram and that you would need a license from the government to buy any.

The plutonium-239 that Oak Ridge is selling is in powder form. Inhaling it would be very dangerous.[2] It is estimated that if you inhaled 20 milligrams you would die of fibrosis in something like a month. Inhaling a milligram would certainly lead to lung cancer. The Department of Energy has set a maximum permissible concentration in air for people who work with plutonium of 32 trillionths of a gram per cubic meter, compared to an inorganic lead (common lead compounds) concentration of 50 millionths of a gram per cubic meter. This aside, if you wanted to made the powdered plutonium into a metal, you would have to repeat many of the steps, with considerable guidance from the open literature, that I have described before. On the bright side, there is the purity of the isotope that Oak Ridge is selling, between 99 and 99.99 percent pure plutonium-239. This would be super weapons–grade plutonium. The Nagasaki bomb used plutonium of almost this purity because it was rushed out of the reactor as soon as enough was ready. Speaking of weapons-grade plutonium, to believe that as far as atomic weapons are concerned, only it, and not reactor-grade plutonium, poses a proliferation threat is to commit the fallacy of reification—the confusion between the name of a thing and the thing itself. To repeat what I noted previously, weapons-grade plutonium can contain no more that 7 percent of plutonium-240. But along with this isotope are several others with much lower concentration. Some examples are instructive. There would be small amounts of plutonium-238, -241, and -242. I would like to comment on these isotopes as well as on plutonium-240.

All of these isotopes are radioactive. Plutonium-238, -239, -240, and -242 are alpha-particle emitters with different half-lives varying from 88 years for plutonium-238 to 376,000 years for plutonium-242. Plutonium-238, -240, and -242 have substantial rates of spontaneous fission, which means they emit neutrons that could pre-detonate an explosive chain reaction. On the other hand—and this is perhaps surprising but important to understand—all of them have larger fission cross sections than uranium-235. In fact, plutonium-238 has a larger fission cross section than plutonium-239.[3] This means that despite the fact that most of these isotopes spontaneously fission, a nuclear weapon could, in principle, be made out of any of them, or any combination, at varying costs and efficiencies.

In connection with plutonium-238, beware: The same radioactivity that makes it a desirable energy source for satellites, also produces a good deal of heat. It is estimated that if one attempted to make a plutonium weapon with any more than 80 percent plutonium-238 it would melt the components. Nonetheless, it is worth noting that the bare critical mass of δ-phase plutonium-238 is only 15 kilograms. On the other hand, the critical mass of plutonium-242 is estimated to be 177 kilograms. Reactor-grade plutonium is a mixture of these fissionable isotopes.

In a typical light-water-cooled reactor that has been allowed to run for some time, the mixture of plutonium isotopes produced is expected to be something like 40 percent plutonium-239, 30 percent plutonium-240, and 15 percent each plutonium-241 and -242.[4] Plutonium-238 would also be present in a lesser amount, and only if the reactor is left to run for a considerable time before the plutonium is extracted. It is produced in a chain in which uranium-238 absorbs a neutron, becoming uranium-239, along with the emission of two neutrons. Uranium-237 beta-decays into neptunium-237, which can absorb another neutron to become neptunium-238, which in turn beta-decays into plutonium-238.

While for evident security reasons, the precise isotopic composition of the reactor-grade plutonium bomb that was tested was not revealed, but what was revealed in 1977 was that in 1962, a bomb using reactor-grade plutonium was successfully tested in the underground Nevada Test Site. The only thing that was revealed about the yield was that it was less than 20 kilotons—approximately Hiroshima or Nagasaki size. One imagines that by "successful" what was meant was that a nuclear explosion had been achieved. A curious aspect of this test was the provenance of the plutonium. The Atomic Energy Act of 1954 prohibits the use of plutonium produced in American commercial reactors for military purposes. The reactor-grade plutonium in this test was provided by the British under the 1958 United States–United Kingdom Mutual Defense Agreement. The implications of being able to use common reactor-grade plutonium for the problem of proliferation are clear.

The Hanford Site and Rocky Flats are avatars for what has happened with plutonium. Both were constructed under the pressures of the exigencies of war, both hot and cold. It is difficult now that the Cold War with the Soviet Union is over, to say nothing of the hot war with Nazi Germany, to put oneself in the mind-set of that era. Two examples of people who were involved with the reactors at Hanford are Eugene Wigner and John Wheeler. In December 1943, when Wigner was at the Met Lab in Chicago, he came to the conclusion that, on the one hand, the Germans were ahead of us in the development of nuclear weapons and, on the other, they knew the location of the Met Lab and were going to bomb it. He moved his family out of the city. When I interviewed Wheeler he said the following:

I had the mistaken idea, based on what happened in World War I, that we would stay out of the war, and it is very unfortunate that I felt like that. If I had been more convinced, as Wigner and Szilard were, that we were going to get into the war I would have pushed harder to begin making the bomb. I figured out that roughly a half million to a million people were being killed a month in the later stages of the war. Every month by which we could have shortened the war would have made a difference of a half million to a million lives, including the life of my own brother. If someone had pushed the project harder at the beginning, what a difference it would have made in the saving of lives.[5]

It was in a crisis emotional context that these two sites were built and operated.

In January of 1943, General Groves commandeered 670 square miles, much of which was farmland, in Washington State on the Columbia River. The people whose land was taken were not told the reason, but they were allowed to harvest one more crop.[6] Groves had an emotional attachment to the river in which he had fished as a boy. Above all, he was determined that no harm would come to the salmon. From 1943 to 1945, the construction project, which involved some 30,000 workers, cost about $350 million. They built 386 miles of highways and 158 miles of railroad track and poured 780,000 cubic yards of concrete. Nothing industrial on this scale had been built before and certainly not in such a very short time. The DuPont Corporation oversaw the enterprise. By 1963, nine pluto-nium production reactors had been built. When the last of them was shut down in January 1987, they had produced 67.4 metric tons[7] (67,400 kilograms) of plutonium, of which 54.5 metric tons were weapons grade. This plutonium would make about 35,000 pits for bombs. In addition, five heavy-water–moderated reactors were built on the Savannah River Site, near Aiken, South Carolina, between 1953 and 1955. Until they were shut down in 1988, they produced 36.1 metric tons of plutonium. Given General Groves's concern about the Columbia River and its salmon, considerable thought was expended as to how to avoid damage to both.

Eight of the production reactors—the last one starting up in 1955—made use of Wigner's design for the cooling system. Water from the Columbia flowed through tubes that passed through the core of the reactor and then returned to the river—"once-through cooling." It was understood that some radioactive isotopes would be produced and that the water would be greatly heated. In fact, when the water left the reactor, its temperature approached 200°F. To deal with these issues, retention pools were constructed into which the effluent water could be put temporarily before it was returned to the river. The pools were designed to hold the effluent water for two to six hours. It was recognized that when it was released, the water would have a temperature higher than that of the river and there would still be a residue of radioactive isotopes that had not decayed. The expectation was that the returned water would, nonetheless, meet acceptable environmental standards. This did not happen for at least two reasons.

Making plutonium during the Cold War was considered so urgent that the time the effluent water spent in retention pools was reduced to as little as 20 minutes.[8] This aside, no one had predicted the sort of radioactivity that was actually produced. In the first place there were chemicals in the cooling water. Some of these chemicals came from the river and some had been added to keep the pipes in the cooling system clean; for example, some 25 to 40 percent of the phosphorus, which after irradiation became the isotope phosphorus-32, was from the cleaning chemicals. Phosphorus-32 has a half-life of 14.3 days and, if one is exposed to it in sufficient amounts, can lead to bone cancer.

To get an idea of the amounts of radiation that were involved, we need to introduce a common unit used to measure radioactivity: the curie. Originally, the curie was the number of disintegrations per second—37 billion—of a gram of radium. But after much negotiation with Madame Curie, it was given the universal definition of 37 billion disintegrations per second of any radioactive isotope. To set some scales, a pound of uranium-238 has 0.00015 curie of radioactivity, while the isotope cobalt-60 has nearly 518,000 curies. Estimates have been made of how many curies of the various radioactive isotopes produced by the Hanford reactors were ultimately released into the river. The estimated amount for phosphorus-32 was some 230,000 curies while, for example, about 6,300,000 curies of neptunium-239 were released. Some of this came from irradiation of the chemicals I just mentioned, but much of it came from the stress on the fuel elements, which increased when the reactors were ramped up to produce more plutonium. The metal coverings of the fuel elements sometimes split, allowing chunks of the radioactive fuel, some weighing up to a pound, to leak out and be flushed into the river. It is said that there were nearly 2,000 such episodes during the lifetime of the eight reactors.

How dangerous was all this to people who swam in the river, drank its water, or ate its fish? The fact is that no one knows for sure. But it is also a fact that while the Hanford reactors were running, it was a policy not to warn people about fishing or swimming or drinking the water downstream from them. One did not want to cause panic.

As disturbing as this may seem, it is nothing compared to what happened on land. The river, after all, if it does not suffer further pollution, would eventually heal itself. There was no way this was going to happen on land and there was, and still is, a concern that groundwater would be affected and that it could leak into the river. This is such an enormous and emotional subject that to do justice to it, if justice can be done, would require another and different book. Here, I give an abbreviated chronology that will convey the general idea.

In 1989, the Department of Energy (DOE) decided to try to clean up 54 million gallons of radioactive waste stored in 177 underground tanks at the Hanford Site, some of which were leaking. The idea, which was abandoned in 1991, was to solidify some of the waste. It was proposed that year to vitrify the waste—turn the tanks into glass. This plan in its original form was abandoned two years later, since it could not treat the waste fast enough. In 1995, DOE decided to privatize the project by contracting British Nuclear Fuels to do it. The contract was canceled five years later, and the Bechtel Company was hired to speed up the vitrification. Bechtel was awarded a $4.3 billion contract, which in 2002 was increased to $5.8 billion as an incentive to complete the project by 2011. In 2005, it was estimated that it could not be completed before 2015, and part of the construction was halted because of concerns about earthquake safety. The present estimate is that the cleanup will cost about $9.65 billion and will require the further man-hour equivalent of 2,300 engineers working full-time for a year. Some progress has been made. The spent nuclear fuel rods have been removed from the retaining ponds where they had been stored, and the radioactive sludge in the ponds is in the process of being cleaned up. General Groves died in 1970, so he is not available for comment.

Rocky Flats (Plate 13) served an entirely different purpose. In 1950, President Truman ordered a crash program to build a hydrogen bomb. Rocky Flats, which is located on 384 acres 16 miles northwest of Denver, was created to manufacture the finished plutonium pits, which were anticipated to be needed as triggers for the then-still-hypothetical hydrogen bomb and above all for the expansion of the fission bomb program.

The finished pits were shipped to the DOE Pantex Facility near Amarillo, Texas, for final assembly. As of July 1994, that facility housed some 6,000 pits. It holds more now, some of which are 33 years old. Pits age from both the inside and the outside (Figure 18). On the outside there is chemical corrosion, and on the inside there is radioactivity. One of the concerns is that these two processes will upset the stability of the alloy of δ-phase plutonium with gallium. This is a major weapons concern in stockpiling these pits, and much research is being devoted to solving the problem. This research is hampered by the fact that the cohort of plutonium experts is also aging and that, since bomb tests have stopped, the results

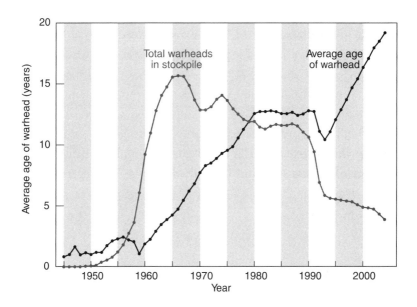

FIGURE 18 Average age of warhead versus year.

must be computer simulated. How certain these computer-simulated results are, I am not sure.

From 1952 to 1975, Rocky Flats was managed by Dow Chemical, but not without serious environmental problems that caused it not to bid to renew its contract in 1974. From 1975 to 1989, when it was closed, it was managed by Rockwell International. Rockwell's tenure at the plant did not end happily. On June 6, 1989, the FBI raided the plant and seized records that purported to show various criminal practices involving negligence and mismanagement. The company eventually pled guilty to 10 counts, including violations of the Clean Water Act, and paid a fine of $18.5 million. That September the site was placed on the Environmental Protection Agency's Superfund list of hazardous waste sites, and in February of 1992, it was transformed entirely into a clean-up site. Until then it was still making warheads for the Trident missile. In March of 1995, the Department of Energy estimated that cleaning up the site would take 70 years and cost $37 billion. It then hired the engineering firm Kaiser-Hill to do the job. Kaiser-Hill had the wisdom to bring in outside consultants, including people from Los Alamos. These people found that the science that had been used previously in planning the clean-up was entirely wrong. It had been assumed that plutonium was soluble in water, which meant that one would have to clean up all the water sources on the site. But this did not take into account the peculiar chemistry of plutonium. It is not soluble in water, but rather it collects in small particles on the ground. This vastly simplified the clean-up process, which now became a matter of soil removal. This process was completed in October of 2005 with the final closure slated for December of 2006 at a total cost of $7 billion. It will take some time to restore the landscape and to make sure that there is no further transport of plutonium.[9] On February 14, 2006, a federal jury, after a 16-year lawsuit, found that Dow Chemical Company and Rockwell International Corporation had contaminated the private land in the neighborhood of Rocky Flats perhaps irreparably. The 13,000 plaintiffs in the class action suit were awarded $553.9 million in damages.

All of the countries that have successfully tested plutonium-based nuclear weapons must have gone through similar steps to the ones we did. They must have had their own Cyril Smiths and Oppenheimers. It would be fascinating to know who these people were. How much did they invent, and how much did they learn from espionage or the open literature?

I can trace the path for three countries: Great Britain, the Soviet Union, and China. They are, in an odd way, all linked. In 1961, the British metallurgist H. M. Finniston, of the Atomic Energy Establishment in Harwell, wrote an article in which he described the British program.[10] I found some of this article quite curious. He notes that just after the war, the British and Americans were no longer allowed to collaborate with each other on nuclear weapons. In fact during the war there were limitations on what the British were allowed to observe. They were not allowed to visit the Hanford Site, for example.[11] The British collaborated with the Canadians. What I found odd in Finniston's account was his explanation of why it took them until 1952, when the British first successfully tested a plutonium weapon, to achieve the same results that had been obtained at Los Alamos. The British had, he wrote, to make all the same mistakes that the Americans had made. But why? There was a British delegation at Los Alamos. Why didn't they tell their countrymen when they got back what the people at Los Alamos had done? In particular, the British had to rediscover for themselves the phases of plutonium, how to alloy it, and what kind of crucible to use to refine the metal. Part of the explanation is that the British had no metallurgists in their delegation at Los Alamos. When they left Los Alamos they were not allowed to take any of the classified documents with them. But there was a member of the British delegation who had a photographic memory and did tell what he had learned. This was the physicist Klaus Fuchs, a German-born crypto-Communist, and he told it to the Russians.

It is important to understand the matrix in which Fuchs's revelations were located. In 1943, the Russian physicist Igor Vasi'evich Kurchatov became the director of the nascent Soviet nuclear weapons project. Note well the date.[12] Kurchatov suggested the use of plutonium for which he used the name ekaosmium. This seems to have been inspired by his reading of the paper of McMillan and Abelson on the discovery of neptunium. But just as he finished his report, he was allowed to see the early material that had been obtained by espionage, primarily from Fuchs. This material persuaded him that plutonium was the answer and that the Russians should immediately embark on a reactor program.

I had always been curious as to what exactly, in the long run, Fuchs revealed. We now know, thanks to the release of a document sent to Lavrenti Beria by a V. N. Merkulov and dated October 1945. The source is not specified, but it could only have been Fuchs who had witnessed the Trinity test and was intimately familiar with the design of the bomb that was tested. Here is what the report says under the rubric "Active Material":

> The element plutonium of delta-phase with specific gravity 15.8 [the density in units of grams per cubic centimeter] is the active material of the atomic bomb. It is made in the shape of a spherical shell [as the report makes clear shortly, what is meant is a solid sphere—the Christy gadget] consisting of two halves, which just like the outer spherule of the initiator [the device that produces the neutrons that start the chain reaction and whose design is given in detail in a previous paragraph], are compressed in a nickel–carbonyl atmosphere [which coats the outer surface with a protective coating of nickel]. The outer diameter of the ball is 80–90 mm. [Millimeters—the nine centimeters that was the diameter of the Christy gadget.] The weight of the active material is 7.3–10.0 kg [kilograms]. Between the hemispheres is a gasket of corrugated gold of thickness 0.1 mm, which protects against penetration of the initiator by high-speed jets moving along the junction plane of the hemispheres of active material. These jets can prematurely activate the initiator.
>
> In one of the hemispheres, there is an opening of diameter 25 mm, which is used to insert the initiator into the centre of the active material, where it is mounted on a special bracket. After inserting the initiator, the opening is closed with a plug, made also of plutonium.[13]

Not much is left to the imagination.

If Igor Vasilyevich Kurchatov was the Oppenheimer of the Soviet project, then A. A. Bochvar was its Cyril Smith. Bochvar, a metallurgist, was given the job of making the pits. First he had to make the metal, which meant that first there had to be reactors that made the plutonium. Kurchatov later emphasized that, although Fuchs's information was very useful, nonetheless they had to carry out the work that implemented it themselves. In fact, it took four years from the time that Fuchs's report was made available until August 29, 1949, when the first Russian atomic bomb was successfully tested. Without Fuchs's information it might have taken a couple of years longer, but it still would have happened. The Chinese bomb program is an amalgam of all of this. In the mid-1950s there was an exchange between the Russians and the Chinese. In return for supplying nuclear information the Chinese would supply the Russians with uranium. There were Russian advisers in China, and the Chinese sent many students to Russia to be trained. The Russians began to help the Chinese build a reactor and a gaseous diffusion plant to separate uranium isotopes. The culmination of all this cooperation in the late 1950s was to have been delivery to the Chinese of a sample bomb along with instructions as to how to make more. But then relations soured. The Chinese were left to build a nuclear weapon on their own. By this time they knew about plutonium and implosion. They made the decision not to try to make a plutonium bomb but rather one that used uranium-235. However, they also decided to ignite it by using implosion, which would significantly improve its characteristics. Less uranium-235 was needed. It was tested successfully on October 16, 1964, in an empty desert lake bed in west central China.[14] The Chinese began producing plutonium in the 1960s.[15] By the time it is thought that the two major reactors stopped producing plutonium in the early 1990s, it was estimated (the Chinese have not given the numbers) that a total of about 2.8 metric tons of weapons-grade plutonium had been produced.

If you look at the global inventory of plutonium, both for civilian power reactors and for military purposes, the outlook is quite discouraging. At the end of 2004 it was estimated that there were about 1,740 metric tons of nonmilitary plutonium. This number is a moving target because the amount increases each year as more plutonium is produced in these reactors. At present, at least 70 tons a year are being produced. In looking over the inventories country by country, some results are what you would expect if you possess some knowledge of how reactors have been used to produce electricity in these countries.[16] To give a few examples, in metric tons:

the United States, 403; Germany, 93; Japan, 152–154; France, 231; and Russia, 126. But there are countries such as Sweden, Belgium, and Spain with only 42, 24, and 27 metric tons, respectively. These are not countries that have been at the forefront of the development of nuclear power. The Russians have collected from places such as Ukraine and Bulgaria most of the civil plutonium that they once possessed. This is counted in the Russian total (Plate 14).

Military plutonium is a fixed target because all of its producers— the North Koreans being a prominent exception—have declared that they are no longer producing plutonium for weapons and the like. Production was suspended in the 1990s. These countries presumably have all that they could conceivably ever need. Globally it is estimated that there are about 155 metric tons of such plutonium. The distribution is also what you would expect: in metric tons as of 2003, the United States, 47; Russia, 95; China, 4; and Israel, 0.56. How much North Korea has has been estimated to be 10 to 50 kilograms.[17] The simple fact is that the world is awash in plutonium, most of which we can do without. The question is, What to do about it?

It is clear, at least to me, that the problem is one of politics and economics, not technology. As far as I can see, most of this excess plutonium will have to be stored. It will have to be agreed that, while no storage facility is perfect, what has been proposed is a lot safer than having the stuff in not very secure locations from which it can be stolen and trafficked. Periodically, some interest is shown in this matter, but by and large, there isn't much interest except to protest if it looks as if it might be stored too close to your backyard. Some of the plutonium can be "burned," that is, used up in power reactors. One promising idea involves using what is known at MOX, which stands for mixed oxides. The mixture here consists of plutonium oxide and uranium oxide. The uranium is natural uranium and the plutonium can be whatever mixture of isotopes you want to burn. The mixture is fabricated into a fuel pellet that has about 7 percent plutonium. The pellets are used as reactor fuel with the plutonium isotopes fissioning along with uranium-235. Since the mixture has so little plutonium and uranium-235, it cannot as it stands be used for a nuclear explosive. Needless to say, to avoid proliferation one would have to make sure that the uranium is not being reprocessed. At the moment, only a few percent of the plutonium produced in power reactors is being burned this way. The process is expensive, and once again it is a matter of will. The plutonium story, as I hope I have convinced you, is full of ironies, not the least of which is that what once cost us millions to produce will now cost us billions to get rid of.

In mid-July of 1939, Eugene Wigner and Leo Szilard drove to Nassau Point on eastern Long Island to speak to Albert Einstein, who was vacationing there. They knew of his long-standing friendship with Elizabeth, the Queen of Belgium. They also knew that the Germans were beginning to work on nuclear energy. They wanted to ask Einstein to write to Queen Elizabeth and urge her to stop shipments of uranium from the Belgian Congo to Germany. They also got the idea of writing to President Roosevelt about the danger of German nuclear weapons. Szilard wrote a draft of the letter that was eventually sent to the President. On this visit, Wigner and Szilard told Einstein how a chain reaction could produce nuclear energy. Einstein was very surprised and told them that this was something he had never thought of. Then he added that this would be the first time mankind got energy that was not directly or indirectly derived from the Sun. He probably did not know then that, like the other heavy elements, uranium and plutonium are created in supernova explosions. Like the traditional Faustian bargain, it came from the heavens and, in the case of plutonium, decayed away. As we have seen, no effort or expense was spared during the war to re-create plutonium for military use. It has almost no other use. Now we are stuck with it. As has often been said, "If you sup with the devil, bring a long spoon."

Notes

Prologue

1. Rainer Karlsch, *Hitler's Bombe,* Deutsche Verlags-Anstalt, Munich, 2005.
2. See Jeremy Bernstein, *Hitler's Uranium Club,* Copernicus, New York, 2001.
3. Karlsch, op. cit., pp. 322–324. In his book the first three pages of the patent, which are extremely interesting, are not quoted. I am grateful to Professor Karlsch for providing the full document and related ones.
4. KWI für Physik, Nr7, Pu, p.3. I am grateful to Pete Zimmerman for help in the translation and for many useful remarks and to Peter Kaus for help with the German.

II The History of Uranium

1. A very nice account of this discovery is given by Bertrand Goldschmidt and can be found at *http://ist-socrates.berkeley.edu/~rochlin/ushist.html.*
2. The term "atomic weight" is often employed for this concept. To a purist physicist like myself this usage leaves something to be desired. Weight is gravitationally related. An atom on the Moon weighs a sixth of what it does on Earth, but it has the same mass. I thank Roald Hoffmann for discussions of this and other matters connected with this book.
3. For this table and for the Sanskrit nomenclature, see Subhash Kak, "The Chemist and the Grammarian: Mendeleev and Sanskrit," available at *www.swaveda.com.*
4. Kak, op. cit.

5. The similarity of this number system to Greek was one of the things that persuaded the eighteenth century linguist William Jones that Greek and Sanskrit had a common origin—Indo-European. I am grateful to Rosane Rocher, who teaches Sanskrit at the University of Pennsylvania, for her comments.
6. Kak, op. cit.
7. I am grateful to Freeman Dyson for pointing this out.

III The Periodic Table

1. I thank Andrew Brown for comments on this.

IV Frau Röntgen's Hand

1. These details can be found in Goldschmidt, op. cit.
2. The text of this paper in translation can be found at *http://web.lemoyne. edu/~giunta/EA?BECQUERELann.HTML.* In a very interesting note, Fathi Habashi, Bull. Hist. Chem. 26, No. 2, 2001, points out that this same observation was made and reported on in 1858 by an amateur French scientist named Niepce de St-Victor. There was no context at all in which to place it at the time. I am grateful to Carey Sublette for supplying this reference and to Professor Habashi of the Université Laval for a copy of his note and for comments.
3. See, for example, Albert Einstein, *The Principle of Relativity,* Dover, New York, 1952, p. 71.

V Close Calls

1. This and other relevant quotations can be found at *http://www. childrenofthemanhattanproject.org/HISTORY/H-02d.htm.*
2. Both of the Joliots's lectures can be found at *www.NobelPrize.org.*
3. Emilio Segrè, *A Mind Always in Motion,* University of California Press, Berkeley, 1993.
4. Enrico Fermi, *Nature,* 133, 898–899, 1934; this can be found at *http:// dbhs.wvusd.k12.ca.us/webdocs/chem-History/Fermi-transuranics-1934.html.*
5. See Marco Fontani, "The Twilight of the Naturally-Occurring Elements: Moldavium(Ml), Squanium(Sq), and Dor (Do)," *http://5ich-portugal.ulu-sofona.pt/uploads/PaperLong-MarcoFontani.doc.* This site contains this and other interesting facts about the search for the missing elements.

6. The entire paper can be found at *http://dbhs.wvusd.k12.ca.us/webdocs/Chem-History/Noddack-1934.html.*

7. See Emilio Segrè, *Enrico Fermi, Physicist,* University of Chicago, Chicago, 1970, p. 80, for this quotation and some of the details of this incident.

8. I thank Richard Garwin for a discussion of this point.

9. Segrè, 1970, op. cit., pp. 85–86.

10. Ibid., p. 86. He calls the pulses "ionization pulses" because the fission fragments when they plow through matter knock off some of the atomic electrons leaving behind positively charged ions that would show up in a detector.

VI Fissions

1. Bernstein, 2001, op. cit., p.143.

2. For an excellent biography of Meitner, see Ruth Lewin Sime, Lise Meitner, University of California Press, Berkeley, 1996. The late Max Perutz wrote an extensive review of the book in the *New York Review of Books*, 44, No. 3, February 20, 1997. Perutz has a kinder view of Hahn's behavior than I do.

3. Sime, op. cit.

4. Ibid., p. 28.

5. Perutz, op. cit.

6. Bernstein, 2001, op. cit., p. 145. In 2004, I paid a visit to Farm Hall and was surprised by how small it is. The 10 detainees were at fairly close quarters. The new owner explained that when he bought it he had no idea of its history. When he had some of the floorboards torn up, he discovered a number of wires that he could not explain. They turned out to be part of the equipment used to record the Germans.

7. Portions of Bagge's diary can be found in Erich Bagge et al., *Von der Uranspaltung bis Calder Hall,* Rohwolt, Hamburg, 1957. This quote can be found on pp. 66–68.

8. Otto Hahn, *A Scientific Autobiography,* Scribners, New York, 1968, pp. 90–91. This is typical of the elliptical self-serving style of this autobiography. No editor seems to have challenged it.

9. Not all neutron interactions obey the $1/v$ law described above. Fission is one that does.

10. Sime, op. cit., gives an excellent detailed account of this in Chapter 8 of her book.

11. Bernstein, 2001, op. cit., p. 142.
12. Hahn and Strassmann, *Naturwissenschaften,* XXVI, 1938, p. 755.
13. Sime, op. cit., p. 178.
14. Ibid., p. 235.
15. I have taken this from Sime, op. cit. She has collated two Frisch sources and in a footnote discusses the dates of the visit. I find it interesting that on this walk in the woods they had paper and some sort of writing implements. But it is a wonderful story.
16. Lise Meitner and Otto Frisch, *Nature,* 143, 239–240, 1939. It is amusing that this paper is signed Meitner and Frisch in that order. This proves once again that the blood of scientific credit is thicker than the water of kinship.
17. Hahn, op. cit., p. 205.
18. E. Crawford et al., *Nature,* 82, 393–395, 1996.
19. C. F. von Weizsäcker, *Nature,* 383, 294, 1996.

VII   Transuranics

1. Hahn, op. cit., *Physical Review,* xi. Seaborg wrote the introduction to the English translation.
2. N. Bohr and J. Wheeler, "The Mechanism of Nuclear Fission," *Physical Review,* 16, 426–450, 1939.
3. Niels Bohr, *Physical Review,* 55, 418, 1939.
4. Emilio Segrè, *Nuclei and Particles,* W. A. Benjamin, Reading, Mass., 1977.
5. Jeremy Bernstein, "The Man Who Sees Past Time," *Johns Hopkins Magazine,* 22–33, 1985.
6. Jeremy Bernstein and Franklin Pollock, *Physica,* 96A, 136–140, 1979.
7. Bernstein, 1985, op. cit., p. 26.
8. An excellent source for the details is Samuel Glasstone and Milton Edlund, *The Elements of Nuclear Reactor Theory,* Van Nostrand, New York, 1952. A lot, of course, has happened since 1952, but the fundamentals have remained the same.
9. This and the part of the Nobel address I am about to quote can be found at *http://nobelprize.org/physics/laureates/1938/.*
10. Ibid.
11. Ibid.
12. Meitner and Frisch, op. cit.

13. Emilio Segrè, *Physical Review,* 55, 1104, 1939.

14. McMillan and Abelson, *Physical Review,* 57, 1185, 1940.

15. Henry D. Smyth, *Atomic Energy for Military Purposes,* Princeton University Press, Princeton, N.J., Nov. 1947.

16. Maria Goeppert Mayer, *Physical Review,* 60, 184, 1941.

17. This is an oversimplification and not the way a modern physicist or physical chemist would describe the situation. In addition to these two forces there is the interaction of the electrons with each other. Since electrons repel each other—they have the same charge—part of the effect of the attractive proton electric force is masked. This has to be taken into account, which Mayer did not. Nonetheless, her conclusion—that for the lanthanides and actinides, it is not the shell of valence electrons that is being filled—is correct. I go into more detail about this when in the next chapter I discuss the very odd chemistry and physics of plutonium. I am grateful to Roald Hoffmann for very helpful comments about this chemistry.

18. Glenn Seaborg, *Adventures in the Atomic Age: From Watts to Washington,* Farrar, Straus and Giroux, New York, 2001.

19. Ibid., p. 127.

VIII   Plutonium Goes to War

1. Bernstein, op. cit., 2001.

2. Louis A. Turner, *Reviews of Modern Physics,* 12, 1, January 1940.

3. Leo Szilard, *Physical Review,* 57, 157, 1940; *Physical Review,* 57, 950, 1941.

4. Louise Turner, *Physical Review,* 59, 366, 1946.

5. I am grateful to Rainer Karlsch for supplying this material and for comments on it.

6. Joliot et al., *Comptes Rendus de l'Académie des Sciences,* Paris 208, 995, 1939.

7. The reference to the popular one is Siegfried Flügge, "Die Ausnutzung der Atomenergie," *Deutsche Allgemeine Zeitung,* 15, August 1939.

8. Nier et al., *Physical Review,* 57, 546, 1940.

9. For a description, see Mark Walker, *German National Socialism and the Quest for Nuclear Power,* Cambridge University Press, Cambridge, UK, 1989.

10. Bernstein, 2001, op. cit., p.140.

11. Ibid., p.163.

12. I have made use of an excellent account of Houtermans's life which can be found in Thomas Powers, *Heisenberg's War,* Knopf, New York, 1993. Remarkably, there does not seem to be a biography.

13. Bernstein, 2001, op. cit., pp. 32–33.

14. Manfred von Ardenne, *Mein Leben für Forschung und Fortschritt,* Nymphenburger, Munich, 1984.

15. See, for example, Pavel V. Oleynikov, "German Scientists in the Soviet Atomic Project," *Nonproliferation Review,* Summer 2000.

16. I am grateful to Rainer Karlsch and Heiko Peterman for showing me this version of the report and for discussions about it.

17. I am very grateful to David Cassidy for providing a copy of this document.

18. The calculation that Peierls did was more sophisticated than this. For an excellent introduction to this calculation and to examine their actual papers, see Robert Serber, *The Los Alamos Primer,* University of California Press, Berkeley, 1992.

19. These critical masses are for the most primitive design and can be reduced with a more sophisticated design.

20. Early in the war, Niels Bohr sent a telegram to England from Denmark that ended by asking that his greetings be sent to "Maud Ray Kent." This was thought to be some cryptic bomb reference until Bohr came to England and explained that Maud Ray had been a governess who took care of his children and that she lived in Kent.

21. Seaborg, 2001, op. cit., p. 96.

IX Los Alamos

1. *The First Weighing of Plutonium,* Atomic Energy Commission Division of Technical Information, Oak Ridge, Tenn., 1967.

2. A good discussion of the relations between the two laboratories and other aspects of how the bomb was made can be found in *Critical Assembly* (edited by Lillian Hoddeson et al.), Cambridge University Press, New York, 1993.

3. This dependence is not too difficult to understand. I have mentioned previously that the radius of the critical sphere should be about the same order of magnitude as the mean free path for fission in, say, uranium. The mean free path is the distance a neutron emitted in the chain reaction travels before it finds another uranium nucleus and causes it to fission.

Under normal conditions, in which the material is not compressed, this is about 11 centimeters for plutonium and about 16.5 centimeters for uranium 235. But if we compress the material, there are more nuclei per cubic centimeter and the mean free path becomes shorter. In fact, it decreases inversely to the density: $r_{av} \sim 1/\text{density}$. If you double the density, the mean free path decreases by a factor of 2. However, the volume of the critical mass varies as the cube of its radius. In the case at hand, the radius is proportional to the mean free path; that is, $V_{crit} \sim r_{av}^3 \sim 1/\text{density}^3$. But we want to know the mass of this critical volume—how much material is needed. This mass is just the critical volume multiplied by the density of the material, $M_{crit} \sim V_{crit} \times \text{density}$. If we now put all of this together, we see that the critical mass decreases as the square of the density. If we double the density, the critical mass decreases by a factor of 4. Therefore, knowing the density is crucial in determining how much mass we need to make a nuclear explosion.

4. The clearest account of these matters that I know can be found in the websites of Carey Sublette, a computer scientist who has made a profound study of nuclear weapons. The two that I use in what follows are *www.nuclearweaponsarchive.org/Nwfaq/Nfaq2-1.html* and the more technical one, *www.nuclearweaponsarchive.org/Nwfaq/Nfaq4-1.html*. The other sites in this series are indispensable for anyone with a serious interest in this subject.

5. For more details see the Carey Sublette website at *www.nuclearweaponarchive.org/Nwfaq/Nfaq8.html*.

6. See *www.arq.lanl.gov/source/orgs/nmt/nmtdo/AQarchive/04summer/zachariasen.html*.

7. *Los Alamos Science,* 151, Summer 1980.

8. This quote can be found in C. S. Smith, "Some Recollections of Metallurgy at Los Alamos, 1943–45," *Journal of Nuclear Materials,* 100, 3, 1981. Smith never wrote an autobiographical memoir. In the histories of Los Alamos, Hoddeson et al., *Critical Assembly,* op. cit., being a notable exception, there is relatively little discussion of metallurgy, although it played a crucial role. Another exception is Edward Hammel, "The Taming of '48'," *Los Alamos Science,* 26, 1, 2000.

9. Smith, op. cit., p. 4. There are some curious things in Smith's account. At one point he refers to hydrogen as an "excellent moderator" for high-energy neutrons. This is true in the sense that, as the lightest element, it takes up the most momentum during a collision. The problem is that

protons capture neutrons, which take them out of the chain reaction cycle. This is why graphite rather than ordinary water is used as a moderator in reactors when the fuel is natural uranium. There are light-water reactors that operate with enriched uranium. The idea of using uranium hydrides for bombs was studied at Los Alamos but, as far as I know, was never realized. In Smith's essay he discusses stabilizing plutonium with alloys. He refers to the γ phase. I think he means the δ phase.

10. *Los Alamos Science,* 23, 164, 1995.

11. Ibid., p. 130.

12. E. S. Makarov, *Crystal Chemistry of Simple Compounds of Uranium, Thorium, Plutonium, Neptunium,* Consultant Bureau Inc., New York, 1959.

13. *Los Alamos Science,* op. cit., p. 152 et seq.; *Human Radiation Studies: Remembering the Early Years,* DOE/EH-0454.

14. Ibid., p. 153.

15. Bernstein, 1985, op. cit., pp. 29–30.

16. Smith, 1981, op. cit., p. 9.

X   Electrons

1. *Los Alamos Science,* 149, Summer 1980. The obituary, from which this was taken, was written by Robert A. Penneman, to whom I am grateful for comments.

2. In an interesting paper by Dulal C. Ghosh and Raka Biswas, *International Journal of Molecular Sciences,* 3, 87–113, 2002, using a plausible approximation method, the atomic radii for 103 elements were calculated and indeed these showed the regularities that had been found empirically.

3. A useful table of these results from hydrogen to molybdenum can be found in Jeremy Bernstein, Paul M. Fishbane, and Stephen Gasiorowicz, *Modern Physics,* Prentice Hall, Upper Saddle River, N.J., 2000, p. 308. There is also a discussion of the theory.

4. I would like to thank Erick Weinberg for pointing this out to me.

5. An excellent account of Rocky Flats can be found in Len Ackland, *Making a Real Killing,* University of New Mexico Press, Albuquerque, 2002. For a more complete discussion of the oxidizing behavior, see S. Herner, *MRS Bulletin,* 26, 9, Sept. 2001, p. 672.

XI Now What?

1. I thank Carey Sublette for this information and for telling me about the Oak Ridge program.
2. This and other details about health risks can be found in *Los Alamos Science*, 26, op. cit., p. 75 et seq.
3. The details can be found in J. Carson Mark, *Reactor-Grade Plutonium's Explosive Properties*, NPT/95, Nuclear Control Institute, Washington, D.C., August 1990.
4. Ibid.
5. Bernstein, 1985, op. cit., p. 9.
6. Much useful information about Hanford can be found in S. L. Sanger, *Working on the Bomb: An Oral History of World War II Hanford*, Continuing Education Press. Portland, Oreg., 1995.
7. A metric ton is a thousand kilograms, or 2,204.6 pounds.
8. Details can be found at the site *www.doh.wa.gov/Hanford/publications/overview/columbia.html*. This is a 2005 publication entitled *Radionuclides in the Columbia River*, put out by the Washington State Department of Health.
9. This history is very well described in David L. Clark, David R. Janecky, and Leonard J. Lane, "Science-Based Cleanup of Rocky Flats," *Physics Today*, September 2006, p. 34. I am grateful to Dr. Janecky and David L. Clark for comments.
10. H. M. Finniston, "Metallurgical Studies on Plutonium in Great Britain," in *The Metal Plutonium*, edited by A. S. Coffinberry, and W. N. Miner, University of Chicago Press, Chicago, Ill., 1961, p. 79 et seq.
11. I am grateful to Lorna Arnold for this information.
12. The standard history of these matters is David Holloway, *Stalin and the Bomb*, Yale University Press, New Haven, Conn., 1994.
13. "On the Origins of the Soviet Nuclear Program," *www.nuclearweaponarchive.org/News/Voprosy2.html*, p. 5.
14. A useful reference is Michael Alter, "A Great Wall for the 20th Century, China's Nuclear Program," *www.georgetown.edu/sfs/programs/stia/students/vol.01/allerm.html*.
15. An excellent source for the Chinese program is David Albright and Corey Hinderstein, "Chinese Military Plutonium and Highly Enriched Uranium Inventories," ISIS, June 30, 2005.

16. These figures are taken from David Albright and Kimberly Kramer, "Civil Plutonium Produced in Power Reactors," ISIS, August 2005.
17. This number is thanks to S. Hecker, personal communication.

Credits

Page 6. Oesper Collections in History of Chemistry, University of Cincinnati.

Page 13 (top and bottom). Lawrence Berkeley National Laboratory.

Page 64. Redrawn by Will Mason, Joseph Henry Press (JHP).

Page 66. Redrawn from a sketch made by Ernest Lawrence.

Page 69. Redrawn by Will Mason, JHP.

Pages 72-73. *www.webelements.com,* © 1993-2007 Mark Winter (The University of Sheffield and WebElements Ltd, UK). All Rights Reserved.

Page 109. Used by permission of William and Zachariah Serber.

Page 115. *Actinide Research Quarterly,* 1st quarter 2005, Los Alamos National Laboratory.

Page 120. Courtesy of *Los Alamos Science,* Los Alamos National Laboratory.

Page 121. Courtesy of *Los Alamos Science,* Los Alamos National Laboratory.

Page 129. Redrawn by Will Mason, JHP.

Page 134. Used by permission of William and Zachariah Serber.

Page 143. Adapted from *www.chemguide.co.uk/atoms/properties/atradius.html.*

Page 148. *www.ncl.ox.ac.uk/it* © S.J. Heyes, Oxford, 1997-1998.

Page 152. Courtesy of *Los Alamos Science,* Los Alamos National Laboratory.

Pages 156-157. *www.ornl.gov/sci/isotopes/catalog/html.*

Page 165. Courtesy of *Los Alamos Science,* Los Alamos National Laboratory.

Photowell.

Plate 1. The Thomas Fisher Rare Book Library, University of Toronto.

Plate 2. History of Medicine, National Library of Medicine, National Institutes of Health.

Plate 3. Stadt Wesel, Archiv, Germany.

Plate 4. AIP Emilio Segrè Visual Archives, Brittle Books Collection.

Plate 5. AIP Emilio Segrè Visual Archives.

Plate 6. Lawrence Berkeley Laboratory, courtesy AIP Emilio Segrè Visual Archives.

Plate 7. Deutsches Historisches Museum, Bildarchiv (Meinshausen Kleinmachnow), Berlin.

Plate 8. University of California, Berkeley, courtesy AIP Emilio Segrè Visual Archives.

Plate 9. Los Alamos National Laboratory.

Plate 10. Los Alamos National Laboratory.

Plate 11. Courtesy of *Los Alamos Science*, Los Alamos National Laboratory.

Plate 12. U.S. Department of Energy.

Plate 13. U.S. Department of Energy.

Plate 14. Courtesy of *Los Alamos Science*, Los Alamos National Laboratory.

Index